ADLAI STEVENSON HIGH SCHOOL
3 3148 00508 6960

D1226351

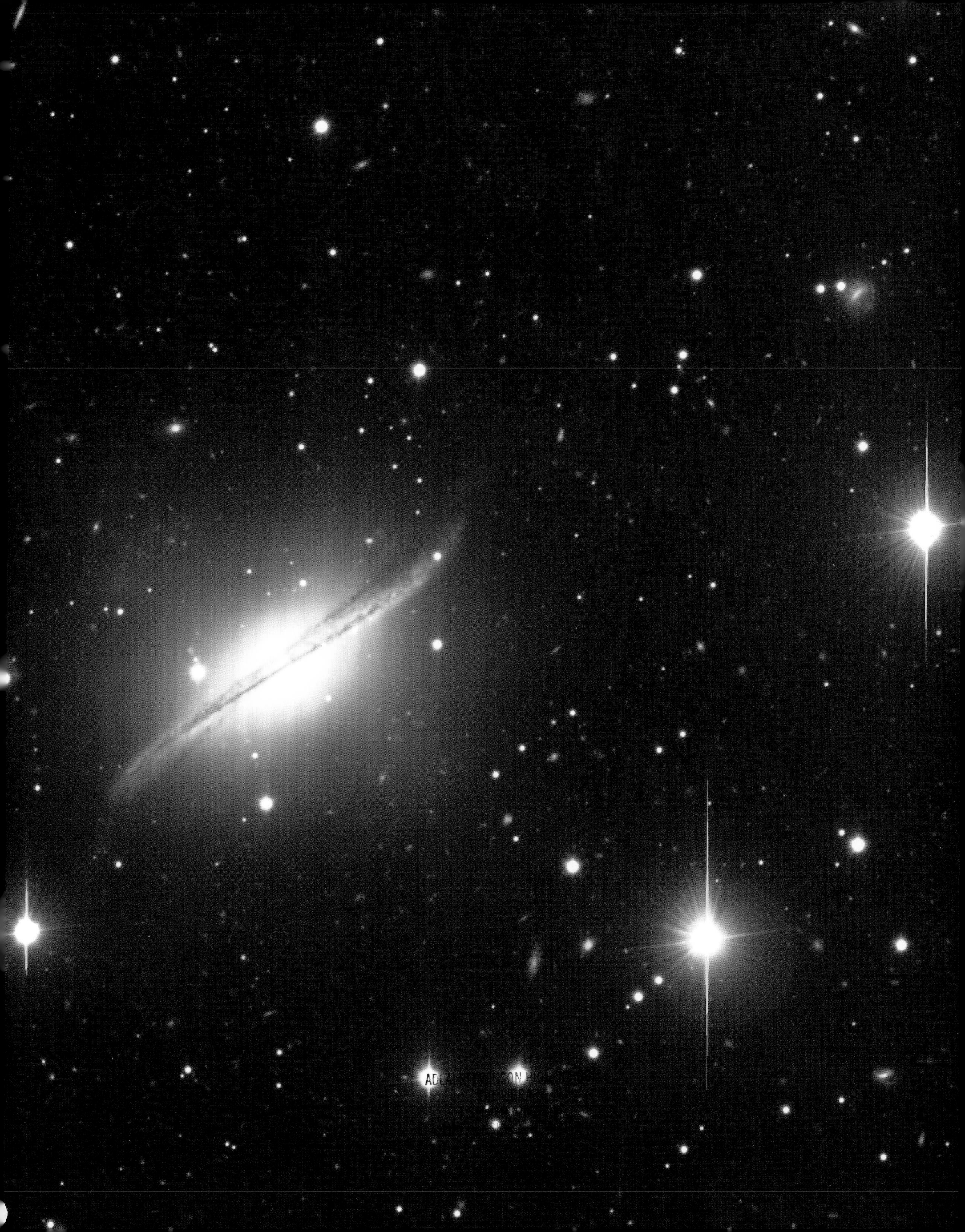

Govert Schilling

Evolving Cosmos

PUBLISHED BY THE PRESS SYNDICATE OF THE UNIVERSITY OF CAMBRIDGE
The Pitt Building, Trumpington Street, Cambridge, United Kingdom

CAMBRIDGE UNIVERSITY PRESS
The Edinburgh Building, Cambridge CB2 2RU, UK
40 West 20th Street, New York, NY 10011-4211, USA
477 Williamstown Road, Port Melbourne, VIC 3207, Australia
Ruiz de Alarcón 13, 28014 Madrid, Spain
Dock House, The Waterfront, Cape Town 8001, South Africa

http://www.cambridge.org

First published in Dutch by Fontaine Uitgevers BV 2003
English edition published by Cambridge University Press 2004

Printed in the United Kingdom at the University Press, Cambridge

Typeface Collis 11/16pt. *System* QuarkXpress® [SE]

A catalogue record for this book is available from the British Library

Library of Congress Cataloguing in Publication data
Schilling, Govert.
Evolving Cosmos / Govert Schilling.
p. cm.
Includes bibliographical references and index.
ISBN 0 521 833.25 6
1. Cosmology – Popular works. I. Title.

QB982.S35 2004
523.1–dc22 2003058671

ISBN 0521 83325 6 hardback

Contents

Foreword

In the hills, far from towns and villages, a shining, silvery dome bestrides a hilltop. It houses a telescope trained on the night sky. All is deathly silent save for the distant sound of a church bell carried fitfully on the wind. The dark, twinkling, impassive sky blankets a silent Earth. Sitting at his telescope, the astronomer reflects on the secrets of the universe.

Is this romantic picture accurate? It is partly accurate but it has changed radically in the last half century. To investigate the Cosmos, astronomers now use big, modern telescopes on remote mountaintops in the Canary Islands, Hawaii or the Andes. And satellites orbit the Earth, focussing on a star, a distant planet, or a galaxy. The results of all of these measurements are processed by astronomers and analysed with powerful computers. Sometimes, years are spent labouring over the results of a few nights, until it is clear what these measurements can teach us about the Universe.

This is the way research has progressed for centuries but now, thanks to the perfecting of instruments and methods of observation, it is going faster and on a higher plane. Researchers have to reflect more deeply, pull out all the stops, and build on the work of their predecessors to find answers to questions that mankind has asked for centuries. Who are we? Where do we come from? What is the Universe? Does it have limits? Does it have an origin? In recent years, research has answered many of these questions and, step by step, we are going further.

Our Earth has not always existed. It was created four-and-a-half billion years ago. The first signs of microscopic life appeared when the Earth was less than a billion years old. We are now more than three billion years further on and are part of the wealth of life-forms that have developed since then. Is the Earth unique in this respect? Are we the only celestial body in the Universe where life can occur? I am absolutely convinced that we are not! Our Universe is ideal for producing life, but we can imagine other universes that are built in such a way that life could never appear.

Other universes? One of the recent developments is that we can justifiably assume that there are other universes besides our own. Just as our Earth is not the only planet, the Sun not the only star, and the Milky Way is not unique, we may also assume that the processes that created our Universe can also lead to the creation of others.

How our Universe came into being and developed is compellingly described in this book. The Universe was born approximately fourteen billion years ago from a miniscule bunching of an immeasurable amount of energy, which gave rise to simple atoms, stars and galaxies. New types of atoms were created by nuclear fusion in stars. Planets were born. In their environment grew increasingly complex molecules and this finally led, in a delicate interaction, to the creation of life.

All of this resulted from an original mass of some of the simplest subatomic ingredients, orchestrated by equally simple laws of physics. The simplicity, the ability to order and the creative power of nature are profoundly impressive.

Evolving Cosmos tells this epic in such a truly brilliant way that I have no hesitation in ranking the book among the best in the world in this field. It shows, in a masterly way, a symbiosis of human thinking, the overwhelming simplicity and power of the laws of physics and – even more so than before, when we saw much but understood less – the profound romance of the beautiful firmament. All of this is excitingly described with great dedication. I congratulate the writer on his achievement and wish the book a good future and a wide international circulation.

Professor Cornelis de Jager,
Emeritus Professor of Astronomy
Utrecht Observatory

1

Creation

The impenetrable Nothing brings forth a seething world of space and time, matter and energy. Here the seed is sown which will grow to become the Cosmos

TO EVERYTHING THERE COMES AN END, AND THIS BOOK IS NO EXCEPTION. Yet this is only the beginning and whatever ends has also once begun. There is no end without an origin, no autumn without spring, no death without birth. Everything around us is in the throes of constant evolution, a state of continual flux. Nothing is as it was or will be what it is; the only constant is change itself. Existence is a capricious stream flowing from the past to the future, from start to finish, from birth to death.

And this law of beginning, growth and end applies even to the all-embracing Cosmos. We cannot fathom its origins, but origins there were. Its end is hidden from sight, but end it will. Yet between these two extremes, every conceivable event takes place: the birth of a star, the division of a cell, the wonder of a child. These are the sparkles on the stream that flows from the Big Bang to the final sigh: fleeting scenes from the life of the Cosmos. And it is a Cosmos that is no bleak, dark emptiness, no terrifying, static vacuum. It is a restless world of wonders; a living, evolving Universe.

You and I are part of this Cosmos. We are inextricably bound up with a Universe that surrounds us on all sides. Our existence is closely interwoven with its evolution. Without the Big Bang, there would be no matter; without matter there would be no stars; without stars there would be no carbon; and without carbon, there would be no life. The DNA in our cell nuclei, the hormones in our bloodstream, and the electrons in our cerebral cortex all obey the same natural laws that govern quasars and pulsars. Indeed, the atomic nuclei in our bodies were once forged in the hearts of distant stars; the quarks and electrons in the world around us still bear traces of the Big Bang. We are one with the Cosmos.

Those who are aware of this know that the past stretches back further than the childhood photo, the history lesson or geology. The birth of an individual is a tiny flicker in the existence of the species; the origins of life, a slight ripple in the history of the planet; the creation of the Earth, a peripheral phenomenon in the evolution of the Universe. The quest for our origins cannot stop at our ancestors, nor at the first vertebrates, and not even at the formation of the Earth. Those who wish to search for our origins must go back to the Big Bang.

That means a radical break with the familiar world around us. Our world may eventually have emanated from this same Big Bang but it bears very little relationship to it. We shall have to travel back through the tunnels of time to the early childhood of the Universe, and further, until all of the past has been used up and the cosmic clock stands at zero. This is a world without stars

Electrically charged particles leave tracks in a bubble chamber. Physicists are using particle accelerators to find out more about the Big Bang.

Was God a mathematician? Unshakeable laws of nature and mathematical precision lay at the foundations of creation.

and galaxies; a Universe without molecules, atoms and nuclear particles; a crazy Cosmos of pure energy. This Cosmos is unbelievably small, unbelievably hot, unbelievably compact and unbelievably young. A quivering point Alpha; a scientific Genesis.

So is this the moment of creation? Not really. Nearly, but not quite. That First Moment lies permanently beyond the view of science – elusive for both observation and theory. No, the clock does not stand at zero, but there is little difference between zero and almost zero. If you take a close look, you will see that 0.000 000 000 000 000 000 000 000 000 000 000 000 000 000 1 second has elapsed. The clock has started; the starting line has been passed. There is no way back. The evolution of the Cosmos has begun.

The complete Cosmos that we can observe through our telescopes, the millions of galaxies with their nebulae, star clusters and planets, the trillions of cubic light-years with all their secrets and unexpected wonders, are all compressed into a seed that is no bigger than an atomic nucleus – a cosmic mustard seed in which the entire future of time and space is imprisoned. Just as Van Gogh's *Sunflowers* already existed in latent form in the daubs of paint on his pallet until they were brought to life on canvas by the hand of the master, so too were the Andromeda galaxy, Kilimanjaro and Mother Theresa hidden in this unsightly primordial atom, awaiting the moment at which the forces of nature would grant them the right to exist.

For the time being, there is no such thing as matter. This shimmering pinprick contains all the energy that there ever was or ever will be. Quarks, atoms and molecules play no part in this first scene. The primordial seed is about to explode – not *in* the Cosmos, but *as* the Cosmos. This is no primordial atom in a limitless, empty space; this is the limitless space itself. The seed *is* the universe; the grain *is* the Cosmos. The raster of the dimensions, the apparently immoveable skeleton of space and time, is twisted, rolled up and compressed here. Parsecs in picometers; infinity in the palm of your hand.

It is not only matter that is still waiting in the wings. The forces of nature, as we know them, have also yet to make their appearance. In this embryonic Universe, they are bunched together, for a brief moment, in an unbelievably powerful superforce that unites the properties of the nuclear forces and electromagnetism. Gravity alone has torn itself loose from this symbiosis at the almost-zero moment at which we see the clock begin to tick. A minimal fraction of an instant later, the other forces of nature follow, and the rule of the superforce has ended.

It is this interplay of interwoven and detaching physical forces that breathes life into the Universe. The process of separation produces such an overwhelming amount of energy that space erupts in a self-sustaining explosion. As if there were antigravity involved, the Cosmos expands faster and faster. Space creates yet more space, filled with the same energy, the same repelling effect. In a chain reaction gone wildly out of hand, the size of the Universe increases exponentially. The grain snaps, the germ bursts apart.

This is the birth-cry of the Cosmos, the moment of ultimate origin, the extreme phase in which time seems to occur on a different scale than that to which we are accustomed. A microsecond seems to last longer than an eternity. In every 000 000 000 000 000 000 000 000 000 000 1 second, the embryonic Universe expands by a factor of two. Two pinpricks, then four, eight, sixteen, thirty-two. After ten factors of two, the Universe is already a thousand times bigger; ten factors of two later and it is a million times larger.

The seed *is* the Universe, the grain *is* the Cosmos

What started as insignificantly small remains within fixed bounds even after such an exponential expansion. The Universe doubles in size a hundred times – more than enough to magnify a speck of dust to the current observable size of the Universe. But in spite of doubling in size a hundred times, the inconceivably small germ from which all of this started is still no bigger than an orange. Its extreme spurt of growth over, the Universe assumes familiar dimensions.

All of this has taken hardly any time at all. A hundred doublings of size, each of which lasted 0.000 000 000 000 000 000 000 000 000 000 1 second. To the casual observer, the clock is still at zero. The experienced observer sees that 0.000 000 000 000 000 000 000 000 01 second has elapsed since the First Moment. And that the first phase of creation has come to an end.

We do not know how the exponential expansion of the Universe happened exactly. Even less do we understand why the cosmic inflationary expansion came to a standstill an instant later. But that this earliest phase in the life of the Universe was of cardinal importance to further evolution, to the formation of galaxies, to the origin of life, is beyond all doubt. Thanks to this inflationary expansion, the early Universe now had the particular properties that were indispensable for its later history.

So what sort of Universe is this that gives rise to the second Genesis? It is small and compact – the billions of cubic light-years of outer space in which we are lost and insignificant are compressed into a couple of decilitres. It is still full of shimmering energy – enough to carry on expanding albeit not at an exponential rate but in a uniform way, as if a racing car has suddenly switched to cruise control after a short, powerful period of acceleration. And the Universe is homogeneous – as smooth and regular as desert sand and as devoid of texture as a table-tennis ball.

And it is also flat. Not two-dimensional like a sheet of paper but flat in a geometrical sense. The three-dimensional co-ordinate grid of length, breadth and height is square and pure. No mysterious space curvature, no intersecting parallels, no hyperbolic mathematics. Every irregularity, every blemish that was present before the inflationary expansion has been blown out, spread out and ironed smooth by exponential doublings, be they positive or negative warps in space-time as a whole or local variations in energy distribution. Every point, every direction is equal in this uniform Universe in which neither position nor orientation appear to have meaning.

What does have meaning, however, is time ticking forward – the cosmic clock that has just started and is still counting in nanoseconds instead of gigayears. Time is change, development and growth. It is growth in the most literal sense in a Universe that is expanding, and in which space is constantly taking up more space. But now that the period of inflationary expansion is over and the primordial energy of creation is spread evenly over the expanding grid of space, this growth is accompanied by an irreversible drop in temperature. From being overheated and super-compact, the Universe is taking its first steps on the path that will inexorably lead to the vacuum and absolute zero.

Never again will the Cosmos be as hot and compact as in this early stage, in which nature

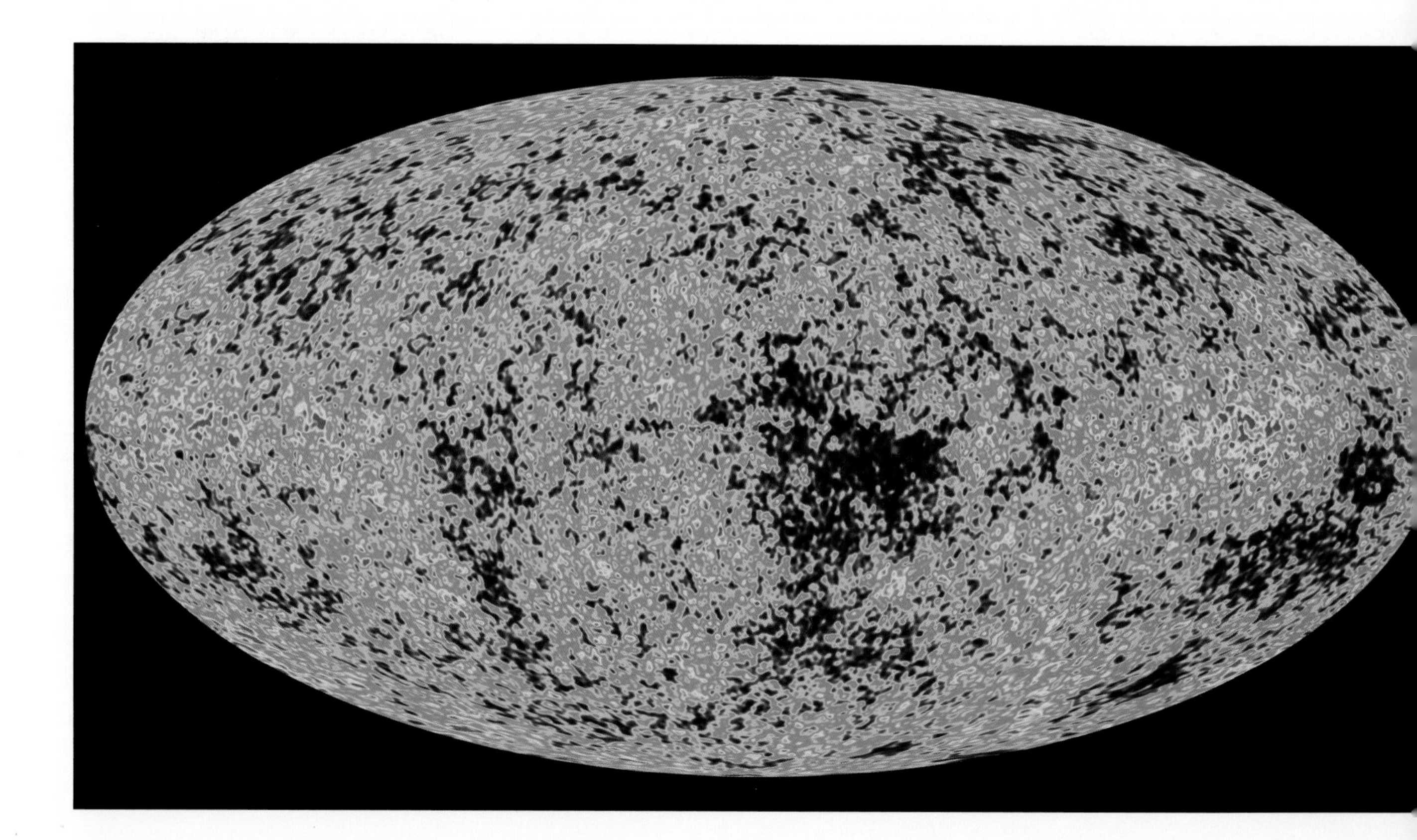

seems to make no distinction between energy and matter. Just as a flower can change its appearance by opening and closing, the contents of the Universe constantly change their appearance also. Radiation and matter, energy and mass are two sides of the same coin, and in the extreme conditions of the origins of the Cosmos, this coin is spinning at amazing speed. Quantum packets of pure energy change into pairs of elementary particles and antiparticles, which then destroy each other in a flash of energy.

Energy and mass are two words for the same idea, two variants of the same variable. Every quantum of energy, every photon of radiation, can transform into a pair of particles; every pair of particles can be transformed into radiation. The greater the energy of the photon, the greater the mass of the resulting particles. The more massive the particles that annihilate each other, the more energy is produced. Nothing is added; nothing is lost. This is nature's immutable law of conservation. Whatever transactions take place, and irrespective of the speed at which this happens, the books must balance to far after the decimal point.

It is in this dance of energy and mass that the tangible Universe will finally be born. Here nature is carving the building blocks from which stars, planets and people will eventually be built. Quarks and electrons, protons and neutrons, atoms and molecules condense one by one from the boiling but gradually cooling primordial soup. The dance lasts only a couple of minutes, but this is more than enough to produce the raw materials for the Cosmos; more than enough to fill the expanding Universe with a hot gas of primordial matter.

The fact that in the end there is something instead of nothing, that the Universe also contains matter as well as radiation, owes itself to a tiny imperfection of beauty on the part of nature – a minuscule asymmetry in the laws that govern the Universe. Whenever energy is converted into mass, into particles and antiparticles, it seems that nature has a very slight preference for what we now call 'ordinary matter'. For every billion particles of matter there are 'only' nine hundred and ninety-nine million, nine hundred and ninety-nine thousand, nine hundred and ninety-nine antiparticles. So when all of these particles and antiparticles annihilate each other and are converted back into radiation, one particle is left over. And this means that every particle of matter in the world around us is an unlikely survivor of the cosmic battlefield that was the Big Bang. Without this subtle 'anti-antimatter policy' of the laws of nature, the Universe would consist of a gradually thinning sea of photons – a cooling bath of radiation in which form and structure would never have come into being. You and I owe our existence to a ten-millionth of a per cent asymmetry.

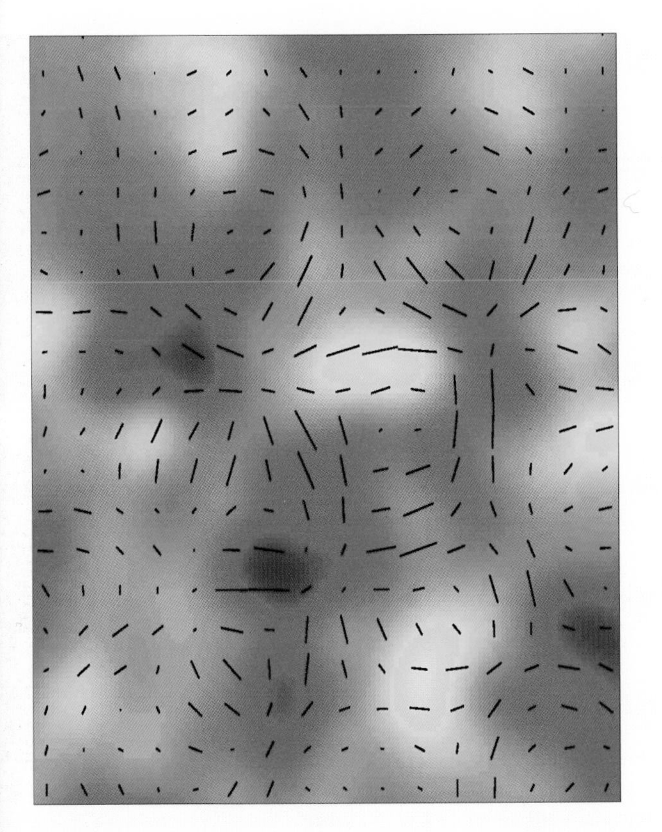

The Big Bang leaves its traces in cosmic background radiation: minimal temperature variations (*left*) and subtle polarization patterns (*right*).

All visible matter in the Universe consists of the minimal fraction of particles that escaped the devilish dance of creation and annihilation. Volcanoes and pulsars, gazelles and gas-nebulae, chestnuts and quasars all consist of the same elements, the same building blocks. Just as a writer can shape an infinity of ideas, meanings and nuances by arranging twenty-six letters in various ways, nature can put together an infinity of molecules, celestial bodies and organisms by meticulously arranging only three fundamental particles.

These three particles are the lead actors in this play: the up-quark, the down-quark and the electron. It is these three that are responsible for the beauty of our night skies, for the violence of tsunamis and typhoons, and for the laughter of a new-born baby. The Cosmos's box of Lego® has only three sorts of block, but their number, and thus the number of possible combinations, is unbelievably huge.

Quarks can't live without each other. And they prefer a ménage à trois. Together, two up-quarks and a down-quark make a proton – a nuclear particle with a positive electrical charge. Two down-quarks and an up-quark make a neutron – practically as large and massive as a proton but with no electrical charge. Protons and neutrons are the ingredients of all atomic nuclei in nature. For example, the nucleus of a hydrogen atom consists of a single proton, while the nucleus of a helium atom consists of two protons and two neutrons. Because of the electrical

You and I owe our existence to a

ten-millionth of a per cent asymmetry

charge of the proton, atomic nuclei also have a positive charge. Only when the atomic nucleus is hidden in a cloud of negatively charged electrons do we have neutral atoms. In their turn, these can join together to form molecules.

Take, for example, the sulphuric acid molecule H_2SO_4. This consists of two hydrogen atoms (H), one sulphur atom (S) and four oxygen atoms (O). A hydrogen atom consists of one proton and one electron. The nucleus of the sulphur atom has sixteen protons and sixteen neutrons. Around the nucleus spin sixteen electrons. Each of the four sulphur atoms has eight protons and eight neutrons in its nucleus, surrounded by a cloud of eight electrons. So the sulphuric-acid molecule consists of fifty protons, forty-eight neutrons and fifty electrons. In Lego® terms, this translates into one hundred and forty-eight up-quarks, one hundred and forty-six down-quarks and fifty electrons. In other words, every molecule in the world around us consists of hundreds of building blocks carefully arranged into unique patterns.

Still other particles were produced in the violence of the Big Bang: neutrinos – maverick particles having no electrical charge and an infinitesimally tiny mass. Neutrinos are not part of the building materials of the Cosmos. They display hardly any interaction with ordinary matter and fly unhindered through fingertips, mountains and planets in countless numbers at almost the speed of light.

And besides an endless sea of criss-crossing neutrinos – more numerous than all of the quarks and electrons put together – the Universe also contains an invisible gas of mysterious

The birth of stars, the creation of planetary systems and the forming of DNA are all phases of the same cosmic evolution.

massive particles of which no one knows the true nature: the unexplained dark matter that is present everywhere and exerts gravitational force but, apart from that, remains as elusive as the Yeti.

These are the offspring of creation, the direct descendants of the Big Bang, the lead roles in the cosmic drama: a thick porridge of energy-rich photons, a sea of elusive neutrinos, a gas of mysterious dark matter and a minimal pollution of up-quarks, down-quarks and electrons. They abide unresistingly by the immutable laws of the Cosmos. Their behaviour is governed by the four fundamental forces of nature and the ubiquitous vagaries of coincidence. There is no script. There is no storyboard. Neither can there be any form of improvisation. Nothing is predictable but there is much that is fixed and unchangeable.

In these first moments, the Universe is hot, compact and non-transparent. The First Moment, the release of the forces of nature and the exponential inflationary expansion, seems to have taken place an eternity ago yet the Cosmos is still less than a second old, its temperature is many millions of degrees and one litre of universe weighs ten thousand metric tons. Radiation and matter cannot pass each other by; they are entangled in a chaotic hip-hop of photons and particles of matter that are constantly spreading and ricocheting in all directions. The Cosmos is an impenetrable cloud of glowing plasma, devoid of pattern or structure.

But the expansion of the Universe generates more space, more elbow room. The plasma cloud expands; the primordial soup is diluted and cools off. Photons lose energy, the temperature drops drastically and it becomes quieter on the dance-floor. The density drops – first to a thousand metric tons per litre, then to a hundred, then ten and then one. Protons and neutrons can finally hold a conversation and enter into a relationship without being instantly scattered by energy-rich photons. The nuclei of helium, deuterium and lithium are formed – small groups of two, three or four nuclear particles.

A couple of minutes after the First Moment, the Cosmos is a nuclear fusion reactor in which new nuclei are being produced. This is where the birth of chemistry takes place, where the

Microcosmos and macrocosmos meet: particle physics examines the building materials of the Universe.

bottom level of the periodic system is being filled up and the chemical composition of the Universe established: 75 per cent hydrogen, almost 25 per cent helium and less than 1 per cent other elements. It is a composition that will change little in the next fourteen billion years of evolution, in spite of the violent nuclear fusion of stars and the production of heavy elements. Carbon, oxygen, sulphur, iron and gold are no more than small pollutants in the cosmic bath of hydrogen and helium that, three minutes after the First Moment, was filled by the mixer tap of the Big Bang.

And while the temperature and density are dropping, and the radiation soup and the sea of particles are becoming thinner, the event clock is starting to tick slower and slower. The Cosmos no longer writes history in nanoseconds. Now the diary of creation is written in minutes, hours, days and years. Many centuries and millenia pass by, tens of thousands of years in which expansion, dilution and cooling have their effect. Finally, the Universe has expanded and cooled

What began as a blinding blanket of visible light has been diluted

to form a plasma whose temperature is a couple of thousand degrees and whose density is less than that of the Earth's atmosphere. Three hundred thousand years have passed – as much time as elapsed between the first *Homo sapiens* and the birth of Albert Einstein.

And now the second and final act in creation also comes to an end. The plasma of electrically charged particles – positively charged nuclei and negatively charged electrons – slowly but surely changes into a neutral gas of hydrogen and helium atoms. The photons in this cooler Universe do not have enough energy to tear an atom apart and thus a mass marriage ceremony takes place between nuclei and electrons. The resulting gas of neutral atoms – a virtually formless ocean of matter – is no longer interwoven with the sea of radiation. Photons are no longer hindered from constantly interacting with charged particles and matter ceases to glow. Radiation is finally able to travel vast distances. The Universe becomes transparent.

✪ This radiation, once as hot and bright as the surface of the Sun, is still present throughout the Cosmos as a constantly cooling and weakening echo of the Big Bang – the cosmic background radiation, older than the light of stars, nebulae and galaxies. In fact, it is the oldest signal in the universe – the original fingerprint of creation. Since the separation of radiation and matter, the Universe has become a couple of thousand times bigger and the waves of background radiation have been stretched a couple of thousand times. What began as a blinding blanket of visible light has been diluted to a barely perceptible noise – a soft whisper at radio wavelengths.

✪ The cosmic background radiation, which comes to us from every direction in the Universe, contains information about the beginnings of cosmic evolution. It is a celestial fossil of the Big Bang – an imprint of the conditions at the end of creation, as if set in stone. And just as the fossil remains of the earliest *Homo sapiens* allow us a glimpse of the origin of man, the background radiation provides us with the subtle ink-blot pattern that lies at the foundations of the present complexity of the Universe. The chart of the background radiation, the baby-photo of the Universe, is, at the same time a *snapshot* of creation, the blueprint of our existence or, as a cosmologist once described it, the face of God.

✪ The marginal temperature variations of the cosmic background radiation – fractions of a thousandth of a degree – show clearly that the primordial soup was not absolutely homogeneous; that there were small variations in density. A little more matter per cubic centimetre here, a little less there. Barely perceptible waves and troughs; the enlargements of quantum fluctuations that date from before the period of inflationary expansion. Without these marginal variations in density, there would be no galaxies, stars, planets and people, and the Cosmos would still be a homogeneous gas of individual atoms, albeit a billion times more rarefied than in the primeval age.

✪ The birth of the Universe is the greatest mystery in astronomy, and no one can look back all the way to the very First Moment. Yet the complex structure of the present Universe can be traced back directly to quantum effects in the first split seconds. The cosmic background

to a barely perceptible noise – a soft whisper at radio wavelengths

radiation provides us with a glimpse of the cradle of the Universe, and in the pink and blue patches on the baby-photo we see the beginning of the majestic architecture of the Cosmos.

2

Contraction

In a Cosmos that is becoming thinner, gravity does its patient work. Two hundred million years after the Big Bang, the first stars ignite in a dark Universe

NOW THAT THE BIG BANG HAS SUBSIDED, the light has been extinguished and the gas neutralised, the Universe remains frighteningly silent and dark for a long time. It is as if the baby has stopped breathing after its first cry of birth. The future of the Cosmos lies entirely in the hands of gravity: if this does not succeed in drawing the expanding matter together again, there will never be a star to shine or a flower to blossom. And at first sight, gravity would seem to have little chance of accomplishing this, for it is by far the weakest of the four forces of nature.

☆ The strong and weak nuclear forces are much stronger than gravity. However, they have a very limited range – lord and master in the nucleus of an atom, but powerless at greater distances. The electromagnetic force is considerably weaker but reaches to infinity. But because there are just as many positively charged particles as negatively charged particles, and because most celestial bodies such as planets and stars are electrically neutral, neither electromagnetism play any meaningful role in the large-scale evolution of the Universe. There remains only gravity, which also has an infinite range but always attracts objects, for the simple reason that there are no particles with a negative mass.

☆ Anyone can easily prove that gravity is far weaker than electromagnetism. A statically charged comb will pull scraps of paper upwards, so the electromagnetic force of the small comb is stronger than the gravity of the entire Earth. However, it is gravity, the weakest actor on the cosmic stage, that will give shape to the Universe and will offer resistance to the expansion of the Universe in the expanding, cooling gas.

☆ Even though, in the far, far distant future, gravity might be capable of bringing the universal expansion to a stop and turning it into a contraction – a possibility that does not seem very likely but cannot be completely excluded – there must be some small-scale condensations if the Universe is not to remain a sea of separate atoms for its entire life. Without subtle variations in density, without the seeds of bunching and structure, the present-day Cosmos would be a dull, dark emptiness – a lifeless vacuum with one hydrogen atom per cubic metre. It would be no wondrous, evolving system but a world of loneliness and silence.

☆ But the seeds are there: subtle differences in density – minute fractions of a thousandth of a gram per cubic metre – like ripples on the water surface presaging the violent storm to come. These are magnified quantum fluctuations from before the period of inflationary expansion that are about to impose their will on the evolution of the Universe. Just as the macroscopic

The end of the darkness: where gas clouds contract, the first generation of stars is ignited.

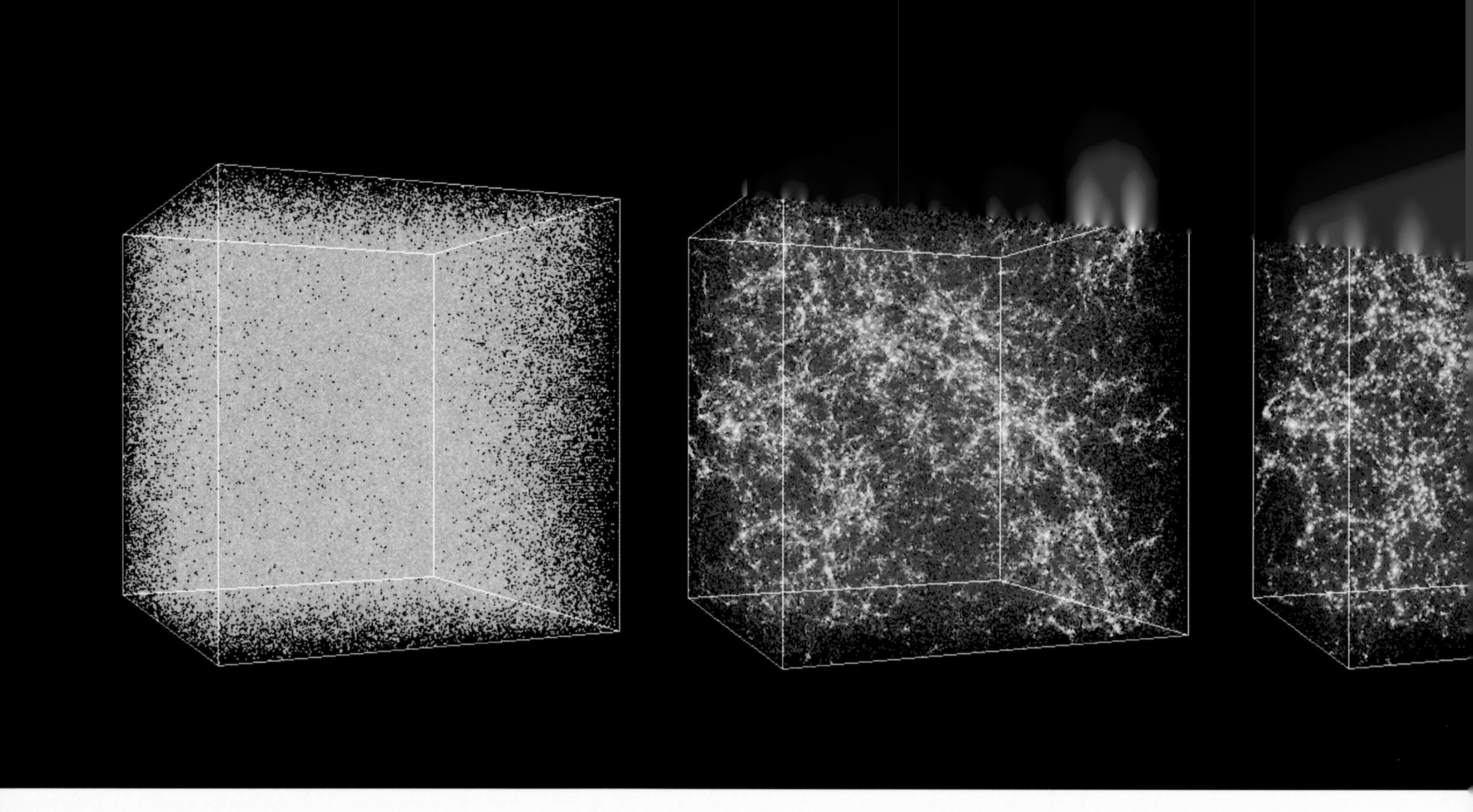

Gas clouds contract under the influence of gravity, creating a cosmic spider's web of dark matter with vast empty spaces.

characteristics of an individual are laid down in the microscopic DNA of the cell nucleus, here too it is the ghostliness of the microcosmos that establishes the foundations for the beauty of the macrocosmos.

☆ One of these primeval quantum shivers, one of the wave-tops in the distribution of matter, develops into the Local Supercluster of which our own Milky Way is a part – a heightened concentration of atoms that will later fragment into galaxies, molecular clouds, nebulae and planets. From one of these wandering pebbles, we look up deep into space and back in time, and

The web of dark matter, arranged in ragged

we see the smudges in the cosmic background radiation that are the silent witnesses of those earlier fluctuations, the blueprint of the cosmic cathedral.

☆ Long before the first galaxies were formed, before the first stars began to shine and the Universe regained its splendour and its glory, the gravitational game plays itself out in the deep darkness of cosmic prehistory. The mysterious dark matter, whose true nature is known to no one but whose total mass vastly outweighs the smattering of hydrogen and helium atoms, slowly but surely begins to coalesce and decides the path that the growth of the Universe will follow.

☆ Where matter is slightly denser than the average, gravity is also slightly stronger. Other neighbouring particles are attracted and gradually mass together. Finally the various clouds of

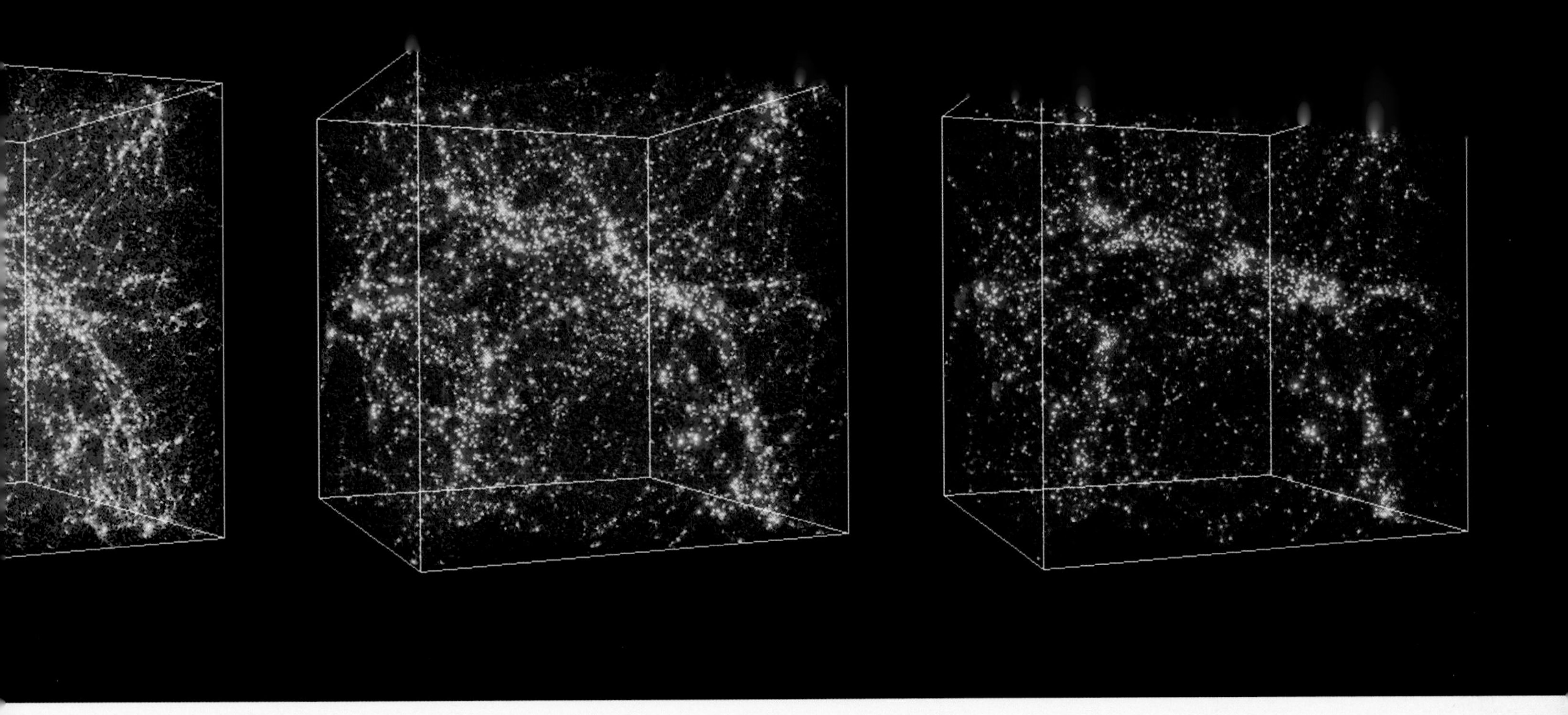

matter begin to feel each other's gravity and start to flow to each other at a higher level in the cosmic hierarchy.

☆ All of this is taking place in an expanding Universe in which matter is being drawn apart by swelling space. It is as if the emptier areas of the Cosmos, where matter is just a bit *less* dense than the average, is being blown up like soap bubbles, in which matter – the soapsuds – is being pressed together. First in thin walls where two bubbles touch each other, then in the seams at which three walls meet, and finally in the intersection of four seams. This is how expansion and

patterns and structures, forms the skeleton of the Cosmos

gravity cause a cosmic schism, an ultimate division between emptiness and fullness, where colossal voids are separated from each other by a remarkable structure of membranes and threads.

☆ The web of dark matter, arranged in ragged patterns and structures, forms the skeleton of the Cosmos. The distribution of atoms in the matter from which the visible world around us is built follows the same pattern of consolidation, like the foam on the wave-tops of an angry ocean. In fact, the present galaxies are the bright tips of dark icebergs, because only where matter is densest can the stars be born that illuminate the Universe.

☆ When the Universe is fourteen billion years old, it will already have more stars than there are grains of sand on the beaches of the earth. But one million years after the Big Bang, it is dark in the Cosmos. Ten million years after the birth of the Universe, the register of births and deaths

Without exception, the first stars in the Universe are large and heavy (artist's impression, *above*). Giant stars are still created now (*left*) but they are much rarer.

is still empty and even after a hundred million years, the first generation of stars is yet to be born.

☆ Where the first star bursts into flame is not known. That its life was short is sure. Perhaps it lived only long enough to witness the birth of the second star, and the third and the fourth, that were extinguished just as quickly to make way for the next five, ten, one hundred. Like cigarette lighters being lit in the audience as the singer on stage performs a ballad, the stars blink on – one by one, first a handful, hesitant and shy, but then more and more until the Cosmos looks like a sparkling fairyland of twinkling lights. After hundreds of millions of years of expanding darkness, gravity has performed its work with the patience of a saint, and the Universe comes to life.

☆ But it's not easy. A star like our Sun consists mostly of hydrogen and helium – the same elements that are readily available in the young Universe – but a thin cloud of gas that contains as many hydrogen and helium atoms as the Sun does not yet turn automatically into a shining star. Although gravity would certainly cause the cloud to shrink, to collapse under its own weight, at the mercy of the mutual gravitational pull of trillions of atoms, it is these atoms themselves that resist the collapse. When the density and temperature increase at the centre of the shrinking cloud, and the atoms collide with each other more often, the influence of the resulting gas pressure makes itself felt. The gas does not allow itself to be compressed further; the gas pressure opposes gravity and the star never forms.

☆ Only when the gas cloud has truly gigantic dimensions and when the number of hydrogen and helium atoms is millions of times greater than that of the Sun, does the gas pressure yield to

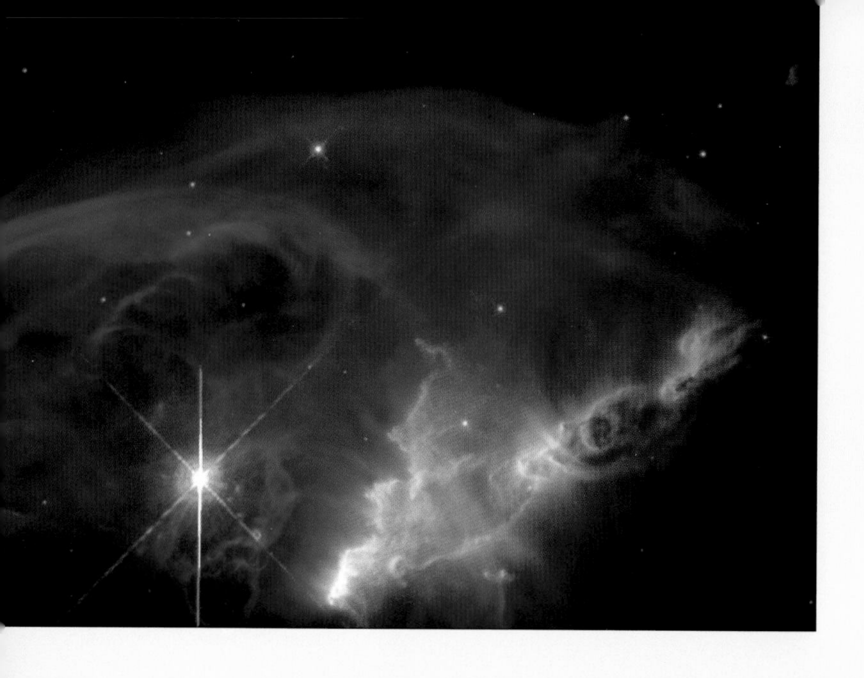

Stars are the superchargers of cosmic evolution. They blast gas bubbles and clouds of matter enriched with new elements into space.

gravity. Only then does a collapse decide the battle. If only there is enough matter available, the high temperature at the centre of the contracting cloud no longer impedes the force of gravity and it collapses inexorably into a star.

☆ So the first stars in the Universe are huge and massive: hundreds or perhaps even thousands of times more massive than our Sun. Superstars, cosmic searchlights – far hotter and brighter than the Christmas-tree lights that will populate the Universe later. The first lightshow in the history of the Universe, the birth of the first generation of stars, was a dazzling spectacle, an overwhelming overture.

☆ The enormous mass of the first stars is also the cause of their short life. Stars derive their energy from the nuclear fusion reactions in their core. The more hydrogen and helium a star contains, the more energy it produces. So it seems logical that a large, heavy star will live much longer than a small dwarf star. But, in practice, it is entirely the opposite. Although the small dwarf star generates little energy, it radiates this energy in small amounts, a bit like a thrifty student who thinks three times before spending every pound and, in so doing, manages to save some money. On the other hand, the giant star is extravagant with its energy budget, like a lottery winner who goes through his millions in a very short time.

☆ In the early days of the Universe, pounds and lotteries are still a long way off but there is still a precious treasure at stake. For the first time since the Big Bang, hundreds of millions of years ago by now, atomic nuclei are once again being forged into heavier elements. Deep in the core of the star, pressure and temperature are as high as in the inferno of creation and nuclear fusion is again taking place. This is not just the conversion of hydrogen into helium but also, at a later stage in the star's evolution, the fusion of helium nuclei into even heavier elements, such as carbon and oxygen.

☆ With the birth of the first star in the Universe, the foundations of life are laid. Never before in the history of the Cosmos have there been stable atomic nuclei with more than a handful of protons and neutrons. The primordial soup consisted of hydrogen and helium with no more than a dash of deuterium and a minuscule hint of lithium and beryllium. There were no atoms from which you could build complex molecules; no elements suitable for a rich, varied chemistry. Without the formation of stars, without the reappearance of nuclear fusion, organic chemistry would not have been possible, self-replicating molecules would be inconceivable and life would be a daydream.

☆ The first star in the Universe is also the first alchemist. The miracle of chemical transfiguration takes place in its innards. Here atomic nuclei are pressed together, neutrons,

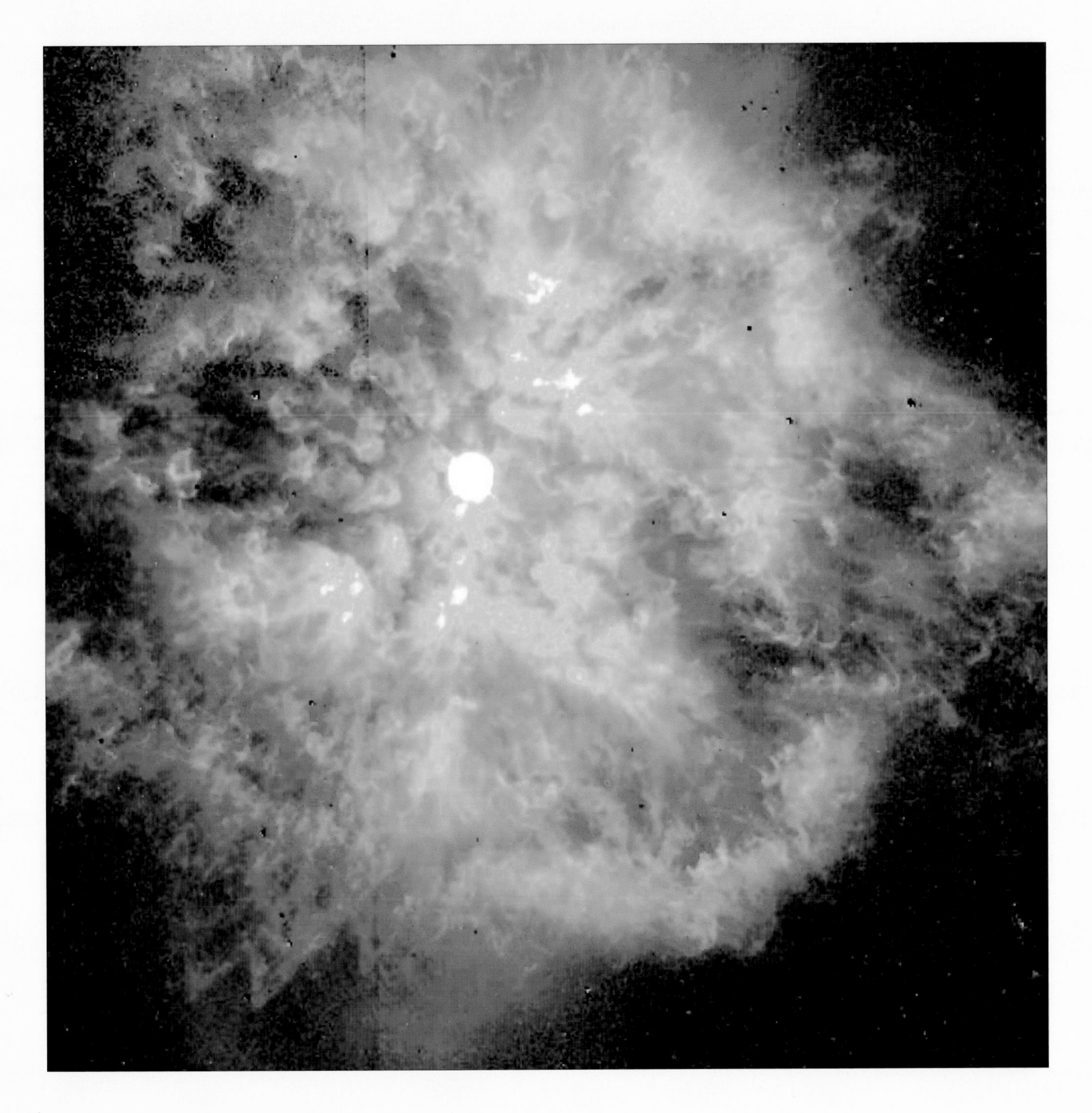

protons and neutrinos become entwined in their quantum dance, and nuclear forces orchestrate the building of new elements. The energy released through this forces its way through the hot, compressed stellar plasma to be radiated into the pitch-black Cosmos on the outside of the giant star.

☆ Now it is the radiation pressure – in fact, the energy produced by the angry young star – that opposes the collapse of the surrounding gas cloud. The star blasts its surrounding area clean; shock waves in the cool hydrogen clouds lead to further compressions and finally to the formation of new stars, just as large and massive and wild.

☆ Stars are cosmic nuclear power stations – celestial factories where raw materials are converted at a fantastic rate into semi-manufactured goods and finished articles. Hydrogen atoms are compressed into helium; helium fuses into carbon; carbon is converted into magnesium and oxygen and later these produce sulphur, silicon and aluminium. Driven by the engine of gravity, the fusion factories produce not only new elements but also colossal quantities of energy. Slowly but surely, they fill up the periodic table and make the young Universe shine.

☆ Never again will it be dark in the Universe; at least not for a long while. Now that this inferno of stars has been kindled, there is no holding back. The energy explosion of the one young star causes something to happen further along the shock waves, from which the next star is born. While gravity kindles new blazes at more and more places, the fusion chain-reaction spreads rapidly along the membranes and tendrils that comprise the cosmic web of coalesced matter. The dark era of the Universe finally comes to an end.

☆ However, the first generation of stars – the Adams and Eves of astrophysics – is not destined

The first star in the Universe is also the first alchemist

to live long. Within a few hundred thousand years, the store of fuel in their cores is exhausted, the nuclear fusion reactor comes to a standstill and gravity once again takes over. Without anyone or anything to save it from its catastrophic fate, the core of the star collapses under its own weight. The forces of nature do their work blindly, and the star exits from the stage, transformed in the blinking of an eye into a black hole – a Gordian knot in space-time, screened from the rest of the Universe by a grim one-way door where quantum gravity stands guard.

☆ But the star does not entirely disappear behind the horizon of the black hole. The pressure and temperature in the imploding plasma are so extreme that the outermost gas layers are hurled into space, carried along on an overwhelming tidal wave of photons and neutrinos. The final breath of the star is a searing hot fireball expanding at the speed of light and destroying everything in its path. The birth cry of the black hole is a titanic explosion that produces more energy in a fraction of a second than our Sun does in ten billion years.

☆ Just as one star was the first-born of the Cosmos, so too is one star the first to die on the battlefield of gravity and nuclear fusion. It is as if nature is discovering its frontiers, discovering and testing new paths and unexpected possibilities – the synthesis of heavy elements, the production of light and heat, the terrors of a black hole, and now the fury of a hypernova: an explosion so powerful that the rest of the Cosmos pales before it and so awesome that it even threatens to topple the Big Bang (now extinguished and almost forgotten) from its pedestal.

☆ The first stellar explosion also marks the start of cosmic environmental pollution. The clean clouds of hydrogen and helium, the pure products of the Big Bang, are contaminated by the combustion products of the stellar nuclear fusion process – the slag and clinker of the nuclear ovens. It is the beginning of a cosmic cycle that will eventually lead to the birth of smaller stars, dark clouds of dust and soot, cool celestial bodies of silicon and iron, people of flesh and blood.

☆ And while the eternal drama of birth, life and death is being played out on the crests of waves

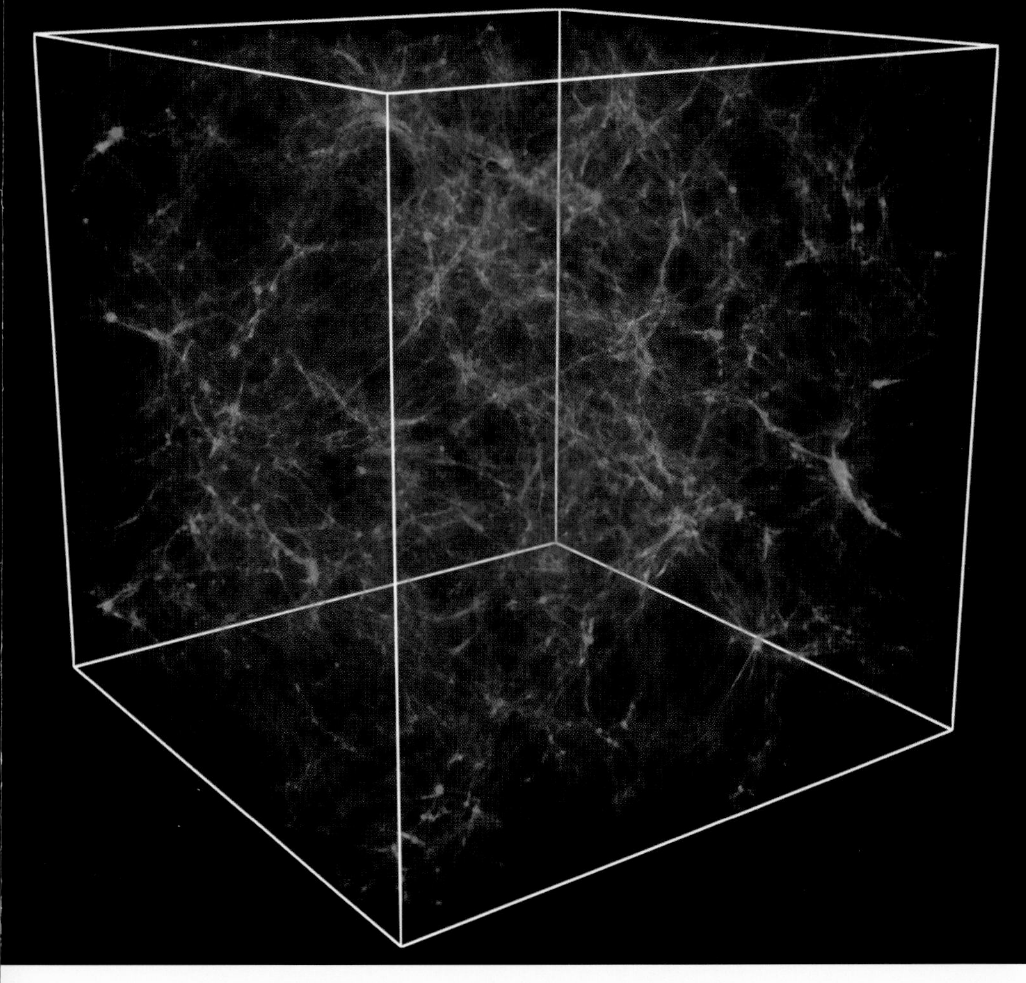

By comparing the two-dimensional distribution of galaxies in the night sky (*above*) with three-dimensional computer simulations (*left*), cosmologists investigate the evolution of the large-scale structure of the Universe.

Overpage:
The gravity of a large cluster of galaxies acts as a lens and distorts the image of objects further away.

Billions of light-years distant, astronomers see small, irregular galaxies that were created shortly after the Big Bang.

Clusters of galaxies are the biggest coherent structures in the Universe – the urban conglomerates of the Cosmos.

of matter, gravity continues to do its work far away from this. The dark matter is continuing to bunch together in an expanding Universe; gas clouds are streaming through the invisible foundations of the cosmic web to the deepest points in the gravitational field and, around the swirling spectacle of stellar birth, nuclear fusion and a terminal explosion gradually heap together more and more matter. Here the foundations are laid for the birth of a complete galaxy – a community of millions or billions of stars.

☆ How the first galaxies came into being is an unsolved mystery, a well-kept secret that the Cosmos does not yet want to give up. Not even the most powerful telescope on Earth or in space can penetrate the vaults of time deep enough to unravel the genesis of these cities and villages of space completely. But although the details may be hidden, the outcome of this hierarchical structuring process is all too evident: from a virtually smooth, expanding ocean of cooling gas, the Universe changes within a few million years to a dark empty sea, dotted with islands of light like glitter on black velvet – an archipelago of small and large collections of stars, organised into groups and clusters.

☆ Possibly each of these protosystems evolves around a short-lived superstar. Perhaps the birth of new stars takes place in gas clouds ravaged by the destructive fireball of a hypernova and polluted by the products of cosmic alchemy, permanently imprisoned in the gravitational pull of

The first stellar explosion also marks the start of cosmic environmental pollution

a lingering black hole. New life in the close vicinity of violent death; glittering, sparkling stars in the shadow of a black hole, like flowers on the slopes of a destructive volcano.

☆ And, imperturbably, gravity continues to play its role: gas clouds collapse into new stars, and protogalaxies – often not much more than irregular groupings of a couple of hundred million stars – gather together into larger objects with more structure and pattern. These are the precursors of the stately galaxies that inhabit the present Universe, each with its own dark secret hidden deep in its innermost core. From now on, it will be the galaxies that play the lead role in

the large-scale evolution of the Universe and which, in fact, must be regarded as the building blocks of the cosmic cathedral. But in these same galaxies, the small-scale life-cycle of stars is playing itself out and the miracle of complexity is taking place at the Cosmos's own pace.

The beginning of a cosmic cycle that will one day lead to people of flesh and blood

3

Ordering

Stately spirals with all-consuming black holes at their cores, the twisted wreckage of cosmic collisions – galaxies are the building blocks of the Universe

The Milky Way galaxy is the home of the Sun and the Earth. It is the residence of life, the homeland of humanity. A slowly revolving disk of hundreds of millions of stars arranged in stately spiral arms wrapped loosely around a brilliant core. At the centre of this core lies a black hole that from time to time emits furious explosions of energy. On the outskirts of the Milky Way, it is quieter. There, cosmic evolution continues on time-scales of millions of years – the birth of new stars, the formation of planetary systems, the slow process of dying of stars like the Sun. And further away from the centre, in the spherical halo that surrounds this flattened galaxy on all sides, old, dim stars drift in a thin bath of dark matter and sometimes seem to almost stand motionless.

✧ But half a billion years after the Big Bang, there is still no trace of the Milky Way. The birth of the Sun and the Earth are still billions of years in the future; the production of atoms from which human DNA will eventually be built has only just begun. The expanding Cosmos is still small and compact and the formation of galaxies has only just started.

✧ These first protogalaxies are irregular clouds of stars, accidental flakes of matter, swept up by gravity and kept together by invisible casings of dark matter. Small and shapeless, they are the building blocks of stately structures such as the Milky Way. Order has yet to be found in these misshapen dwarfs. In unpredictable places, thin clouds of gas collapse into concentrated globules, which in turn give rise to new, massive stars. Young, fiercely radiating suns, hot and blue, set their surroundings ablaze, blow holes in the encircling gas and spew out dust and smoke at the end of their short lives. New black holes – the remains of extinguished giant stars – sink to the core of the protogalaxy and melt into the bottomless pit of gravity that lies there already.

✧ And it is not only the processes in the protogalaxy itself that are causing upheaval. In this very early period, in which cosmic expansion has only just begun, distances are small and next-door neighbours live very close. While every system becomes more massive as it sucks in more of the surrounding cosmic gas, the systems also feel each other's gravity. They whirl around each other in a complex dance; the smallest star clouds are ripped apart by the gravitational pull of their bigger brothers; double and multiple systems arise and formless heaps slowly but surely coalesce into larger structures, into the dignified forerunners of galaxies such as our own.

Billions of stars and ragged dust clouds are packed together in the core of the Milky Way.

Stately spiral arms enfold the core of a galaxy that closely resembles our own Milky Way.

✧ Through the merging of small protogalaxies, not only does the quantity of gas and stars increase but also the mass of the black hole in the centre becomes increasingly greater. What once began as a black hole a couple of times more massive than the Sun, grows into a heavyweight of a hundred, thousand, million times the Sun's mass. As far as the outer perimeter, the concentrated gravity of this monster imposes its will on the motions of stars and gas clouds. Accelerated to enormous speeds, they circle around the massive, dark centre in a vain attempt to escape the all-destroying suction of the black hole. But the fate of most of the matter in the immediate vicinity of the black hole is already sealed: sooner or later, it will disappear forever beyond the horizon of gravity.

✧ This vanishing act does not go unnoticed. Although a black hole does not transmit light and radiation itself, nearby matter does. Just as an emptying bathtub creates a vortex around the plughole, the gas spiralling inwards accumulates into an eddying disk before it is finally sucked in. The pressure and temperature in the disk increase so much that the gas spontaneously starts to produce energetic radiation. This hides the black hole in a radiant envelope that shines brighter than billions of suns.

✧ The radiation of this accretion disk can be so powerful that the attraction of new matter is obstructed. The radiation pressure – the combined energy of all of these seething photons –

Globular star clusters are the oldest objects in the Universe.

A black hole in the core of a galaxy produces a beam of fast-moving particles.

opposes the gravity of the black hole and keeps cooler gas at a greater distance from a disastrous fall into the depths, just as the heat of a fire will keep scraps of paper dancing in the air. The black hole swallows less than it wants; its greed is tempered by the heat of its meal.

✧ Yet more processes are taking place in the cores of the first active galaxies. Besides gravity and radiation pressure, magnetism also plays a disruptive role. The weak magnetic fields in the thin insterstellar gas clouds are mixed up and distorted by the gravitational dive of matter. Where gas is pressed together, the magnetic field lines are closer together so that the strength of the field increases enormously. And where the gas whirls around, the field lines wind up like thread on a spool.

Caught up in a dance of gravity and distorted by gravitational forces, these two spirals are merging into a single, elliptical galaxy.

✧ The motion of electrically charged particles – atomic nuclei and free electrons – is influenced, directed and established much more by the chaotic jumble of invisible magnetic field lines, continuously moving and re-ordering themselves, than by the gravity of the black hole. Electromagnetic forces whip up matter to crazy speeds in unexpected directions; snapping field lines lead to explosive eruptions and, in the core of the galaxy, gravity, gas pressure and magnetism fight a bitter battle for the mastery of matter.

✧ All of this causes the core of the galaxy to blast two piercing jets of gas into space – two powerful beams of protons and electrons at right angles to the rotating accretion disk, precisely along the axis of rotation. Only in these two directions can the gas stream away unhindered at speeds that are virtually the same as the speed of light. In fact, the super-massive black hole acts as a cosmic particle accelerator – a gigantic engine in which gravitational energy is converted, via complicated deviations, into movement and speed.

✧ Hundreds of thousands of light-years from the core, when they have radiated some of their energy, the particle jets are finally slowed down, and even reduced to a stop, by the thin matter in the space between the galaxies. Like the jet of water from a garden hose when it's pointed against the wind, the gaseous beams lose their power and are 'blown about' into colossal, thin clouds that are no longer hot enough to produce X-rays or visible light but that can be clearly seen on radio wavelengths.

Chaos, activity and turmoil are now the norm

✧ So here are the most important elements of an active galaxy: a super-massive black hole surrounded by a glowing-hot, rotating, accretion disk from which two jets of matter are being blown off, which discharge far away from the galaxy into colossal radio-emitting clouds.

✧ In the young Universe, where the formation of the first galaxies has just begun, there are a host of active systems. Black holes merge together and increase in size and mass; 'food' in the form of incoming gas is widely available. Chaos, activity and turmoil are the norm now that the Cosmos is still in its turbulent infancy; only later will peace and regularity make their appearance.

✧ If you were to look around you in this early Universe, you would see active systems in a multitude of forms. If you glanced straight into one of the particle jets, you would be blinded by the radiation of the onrushing gas, enormously strengthened by the relativistic Doppler Effect, caused because the gas is approaching you at almost the speed of light. From a great distance, these 'blazars' look like bright, blue stars that display quickly changing, radical variations in brightness.

✧ If you were to look into the system from an oblique angle, the core would be slightly less dazzling but still extremely bright. In many cases, you would also observe radio waves, either from the core itself or from the clouds at the extremities of the expelled particle jets. These bright radio beacons are called quasars. And if you were to look at the system from the side,

Explosive processes are visible in this X-ray photograph of the core of the Milky Way.

New stars ignite in a circle of light in the centre of a spiral galaxy, probably as the result of an earlier collision.

the bright core would be hidden behind thick, dark clouds of dust and only the two opposing particle jets and the radio clouds would be visible. Blazars, quasars and radio galaxies are the different faces of one and the same cosmic monster: a gluttonous heavyweight hiding in a glittery world of stars, enclosed in a belt of glowing gas, and whose convulsions and belches are visible and tangible at a far distance.

✧ Only later, when the monster's food supply runs low, when the quasar core has sucked its immediate surroundings dry, the accretion disk has disappeared, the particle jets have dried up and the radio clouds have evaporated, only then does the galaxy come to rest, acquire form, and begin to look something like our own Milky Way – a stately spiral of stars, nebulae and star clusters with a heart of darkness, as peaceful and imposing as a dormant volcano that can become active at any moment.

✧ It is in such a galaxy that cosmic evolution advances. Gas clouds coalesce, star clusters split apart, new elements come into being in the cosmic flasks and retorts of astrophysics. Explosions blow carbon, nitrogen and oxygen around; increasingly more complex molecules form in dark clouds and, eventually, somewhere in this system, a molecule will copy itself, a cell will split, a heart will beat. Every process is taking place in this galactic ecosystem, gliding like an island of light through the black ocean of the Cosmos – an independent, miniature Cosmos, a Dinky® toy Universe.

✧ But even in the infinity of the expanding Universe, no galaxy is totally isolated from its surroundings. Other systems pass by like ships in the night, often at a safe distance but sometimes frighteningly close. And, from time to time, there is a catastrophic collision – a cosmic crash that roughly disturbs the relative peace and quiet of space. Not that the three-dimensional motorway is really busy, but there are no halt signs, rights-of-way or traffic lights and gravity is behind the wheel, which means that as a calamity approaches there is no braking or evasive action. In fact, there is purposeful acceleration, so the Cosmos more closely resembles a dodgem ride than a road-safety training ground.

✧ Two large galaxies bound on a collision course notice each other's presence well in advance. The approaching galaxies are not rigid objects like tankers or trucks but mobile

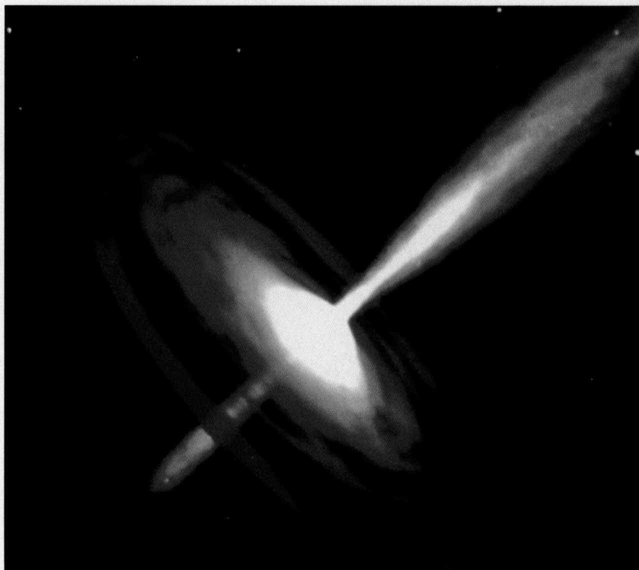

clouds of stars, gas masses and nebulae held together by the same gravity that rules the interaction between both of them. At the 'front' of the two systems, where the distance between the two is smallest and the gravitational pull is strongest, stars loosen themselves from the gravitational hold of the mother system to fall towards each other at increasing speed. The systems begin to lose their well-ordered structure and disintegrate at the edges, like armies on a battlefield when the frontline soldiers advance to do battle for the first time. Spiral arms are unfolded and stretched by the gravitational forces; stars are scattered like splinters of glass in a traffic accident; clouds of gas and dust begin to churn, creating compressions and shock waves.

✧ Deformed and distorted, and surrounded by clouds and strings of glittering debris, the two systems finally slide into each other in a slow-motion process that takes many millions of years. However, there is no fatal crash. The galaxies do not come to a stop but move through each other like ghost ships, to continue their race through space even more badly battered, leaving behind a tell-tale trail of destruction. In fact, a galaxy consists mainly of empty space, and the stars it contains are so far apart that the two systems move through each other like two swarms of mosquitoes without actually touching. In both systems, the orbits of the stars are seriously disrupted by the forces unleashed by this cosmic encounter, but collisions between stars do not or hardly ever occur.

✧ However, what applies to the stars does not apply to the interstellar matter in the galaxies. The tenuous gas and dust clouds *do* collide with each other, with all the consequences one might expect. The gas is heated; shock waves propagate at high speed and, at a host of places, the density increases so much that gravity gains the upper hand, resulting in waves of star formation that wash throughout the galaxies. The fire of star-formation is rekindled not only in the systems themselves but also in the whirling tidal tails. Everywhere, the dark wreckage left by the cosmic collision is illuminated by the bright blue light of young star-forming regions.

<< Two black holes, each with its own accretion disk, approach each other in a merging of galaxies.

< A quasar blasts matter away in two beams coinciding with the angle of rotation of the central black hole.

String-like filaments of gas and baby booms of new stars are the tangible results of a catastrophic explosion in the centre of a misshapen galaxy.

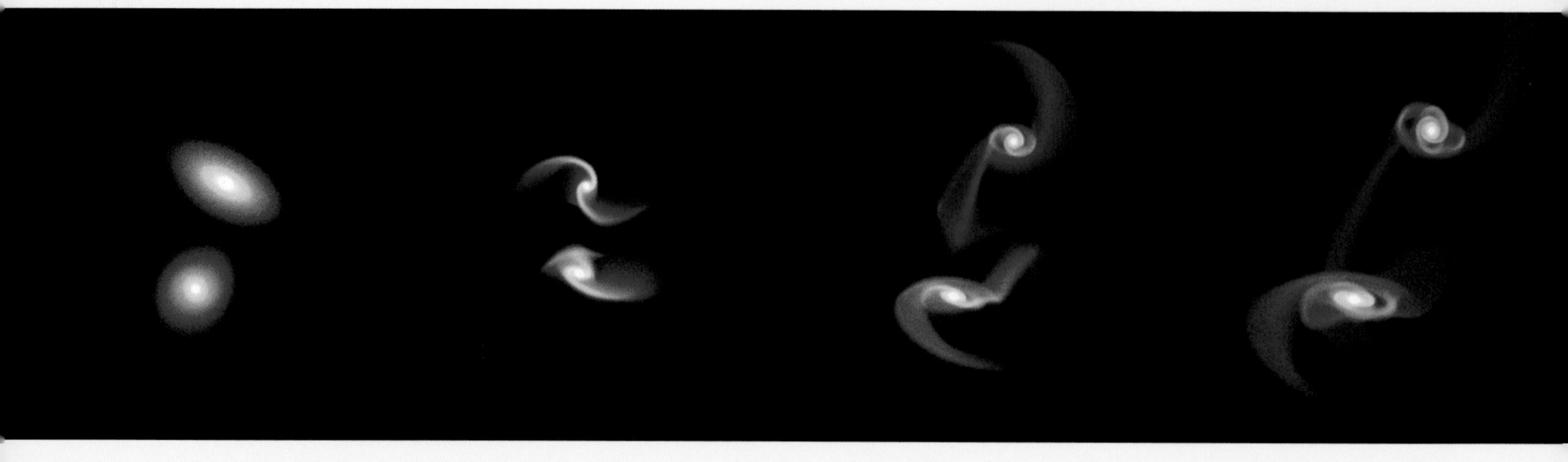

✧ The black holes in the cores of the two galaxies also do their bit. They are rudely awoken from their sleep by the sudden violence of the collision. The dynamic chaos unleashed by gravity stretches its tentacles deep into the heart of the systems. Stars knocked out of their orbits fall prey to the gravitational pull of the black hole; gas that has been pushed aside is sucked inwards. The monster is fed again, the accretion disk recovers; the core of the system spews forth electrons, radio waves and X-rays. The actual encounter is over but the distorted system shakes and shivers for long afterwards, shimmering with energy and permanently scarred by the violence of the collision.

The cosmic meltdown leads to celestial fireworks

✧ Elsewhere, where galaxies collide with a lower relative speed, the spectacle unfurls quite differently. The already slow speed is decreased even further by the friction of interstellar gas clouds. The systems are slowed down, kinetic energy is converted into heat, new stars are kindled, but then the two systems are no longer capable of escaping from each other's gravitational pull. For a moment, they recede from each other, but their fate is sealed: they fall back together, swing back and forth in each other's gravitational field and then finally fuse together into one giant system.

✧ Such a cosmic fusion leads to celestial fireworks. Here too the greed of the black holes in the cores of the two galaxies is intensified and their gluttonous behaviour shows itself in piercing jets of particles and all-penetrating radiation. But it doesn't stop there. The two black holes move ever closer to each other, are whipped into a frenzied circle-dance, spew random fountains of matter and energy into space, and finally merge into one super-massive gravitational monster. The new giant galaxy is ablaze; activity in its massive heart is more violent than ever and, to its furthest perimeters, the cosmic darkness is dispelled by the boiling light of the quasar core.

✧ Little remains of the original shape of the two fused galaxies. They are stripped and dismantled; their spiral arms are ripped apart; their stars are derailed. Their ordered

Two galaxies merge with each other in a computer simulation after being deformed by gravitational forces.

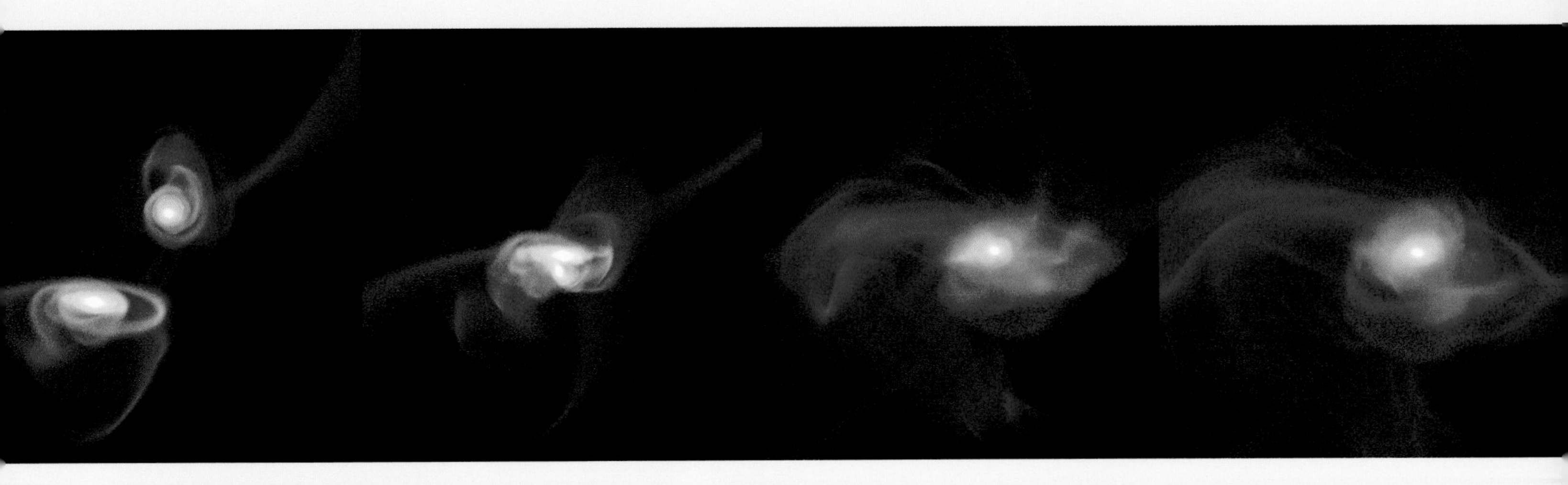

structure has disappeared, as too have their regular appearance and their symmetry. The orphaned stars criss-cross each other, searching for a new equilibrium in a larger orphanage. If they are left in peace for long enough, perhaps this equilibrium is found and they develop into a new spiral structure, more majestic and more imposing than before. But usually, it doesn't get as far as that. Usually, the new galaxy – with its increased gravity – will shortly attract and swallow up other galaxies, create new victims. This finally results in an elliptical system: a gigantic collection of stars swarming around the central core in the most various of orbits. Sometimes roughly spherical, but often rather flattened like a pumpkin or slightly stretched like a rugby ball. In contrast to stately spiral galaxies such as our own Milky Way, an elliptical galaxy does not rotate regularly in a flat plane, with its stars circling around the core like Muslim pilgrims around the Ka'aba, but in a chaotic, three-dimensional form that recalls a swarm of wasps.

✧ Few galaxies escape collisions. If they do not collide front-on with another large system and carry on their journey through the Cosmos misshapen and battered, or lose their identity by merging with a slow fellow victim, then they swallow up smaller surrounding systems – shapeless dwarf galaxies, mini-spirals or small elliptical groups of stars. Sometimes the intruders' footprints remain visible for a long time in the form of ripped-apart strings of stars of various colours, double cores with perhaps twin black holes, or immigrant star populations in the centre that move diametrically opposite to the normal sense of rotation of the galaxy. No single galaxy is autonomous; interaction is the norm. Nothing and nobody withdraws from the evolution of the all-embracing Cosmos.

✧ This is the way the Universe develops. Once it was an almost homogeneous ocean of burning hot gas. Now it is a varied natural-history cabinet of galaxies in the most diverse of forms and sizes: clumsy, pompous elliptical giants; graceful spirals, sometimes as tight as a spring, sometimes as loose as a garden sprinkler; strange lens and ring galaxies; irregular dwarfs and dark ghost galaxies. All these gravitational sculptures are arranged into groups, clusters and superclusters. Each has its own identity but constantly interacts like the multiracial population of a major city. Galaxies are the richly chequered inhabitants of the Universe; the morphologically-diverse building blocks of the Cosmos.

Not all spiral galaxies hide a dark secret in their cores (*left*) but X-ray radiation from the centre of our own Milky Way indicates the presence of a heavy black hole (*centre*). An encounter between two galaxies can also be an optical illusion (*right*).

✧ Our Milky Way is one of these building blocks: a large spiral galaxy on the periphery of a gigantic cluster. An elegantly rotating disk of twinkling young stars, glowing gas nebulae and dark dust clouds; a flattened, stretched accumulation of old stars in the core, arranged around a slumbering black hole; an extended halo of dark matter and orbiting globular clusters. This is the home of white dwarfs and red giants, pulsars and micro-quasars, of supernova explosions and planetary nebulae.

✧ And here, in the suburbs of the galaxy, the Sun is born. Here particles of dust coalesce into planets, organic molecules settle on the hot surface of the young Earth, the wind blows through grasses and ferns, and the reader turns over a page. But not quite yet. The world around you – the pigment in the ink of this book, the oxygen you breathe, the carbon atoms in your body – has yet to come into existence. These atoms have yet to be forged. The cosmic cycle is just starting.

4

In the dark catacombs of cosmic gas clouds sound the birth-cries of new stars. A multicultural society of blue-white super giants and red dwarfs is created here

BIRTH IS A MIRACULOUS PROCESS. It is awe-inspiring but also slow and painful. New life is not self-evident and does not appear unnoticed. For a very brief time, every birth is the centre of the Cosmos – an intense, emotional crossing from safety to poignant vulnerability, a curious mixture of quiet tenderness and bloody screaming. Every new-born baby is a being that is full of unknown promises and unexpected possibilities, like a flower that is beginning to open up but is still a long way from showing its full beauty.

☆ Just as great a miracle is taking place in the cosmic delivery rooms of the Milky Way galaxy, with the same intensity and comparable potency. Stars have no hearts and souls; they do not breathe, they do not think and they do not reproduce. Yet they do know beginning and end, appearance and disappearance, life and death. Like the birth of a child, the birth of a star is accompanied by violence and screaming, at the tedious pace of the macrocosmos. But finally there is a sprouting bud – new life.

☆ New stars have already been seeing the light of day since cosmic prehistory. Long before the formation of the Milky Way, the birth of boiling suns is taking place elsewhere. Deep within them, nuclear forces are preparing the elemental mix from which later, much later, real flesh and blood will be constructed. Everywhere in the Cosmos, nature is working on the enrichment of matter and the increase of chemical versatility. Billions of stars are tirelessly producing huge stocks of carbon, oxygen and silicon day-in, day-out. Smoking giants and explosive supernovae scatter this precious ash over the black sea of the Universe, as building materials for new stars, new worlds, and new life.

☆ And apart from new elements, the aftermath of the life of a star also results in new compounds – molecules of strung-together atoms. There are simple ones like molecular hydrogen and carbon monoxide, but also complex compounds of three, five or ten atoms, macromolecules with carbon chains and rings, small soot particles, silicon crystals and grains of dust. The ash pan of a star is largely emptied into space; the cooled combustion products of hundreds of millions of years of nuclear fusion eddy around the Cosmos. There they mix with the tenuous gas clouds in interstellar space – a slight pollution of cosmic primordial matter.

☆ It isn't much: no more than one minuscule particle of dust in twenty cubic metres. Cosmic dust clouds are cleaner than the most advanced clean rooms on earth, but their inconceivably low density is compensated for by their equally inconceivable dimensions. Just as the light of the setting Sun turns red and weakens because it must travel a great distance through the

Young stars ignite in a multicoloured nebula of gas and dust.

Star-forming close to home: the Orion Nebula is only sixteen hundred light-years away.

Earth's dusty atmosphere, so the light of distant stars changes colour and weakens when it passes through these vast expanses of cloud. In many cases, the dust clouds even block our view of distant stars and nebulae completely and appear as jet-black patches on close-ups of the night sky.

☆ It is in these colossal clouds – sometimes showing up dark against a backdrop of bright stars but often as invisible as a crow in the night – that the miracle of stellar birth takes place. The clouds of cool gas, molecules and cosmic dust are dark cocoons into which practically no radiation penetrates; hidden delivery rooms for new light and life. These molecular clouds are tens of light-years in size and are the biggest coherent objects in a galaxy.

☆ At first sight, they seem to be little different from the clouds of pure hydrogen and helium gas from which the first generation of stars in the Universe were formed billions of years ago. Also then, accidental condensations collapsed under their own weight and hot protostars came into being which, as soon as their cores were hot enough for nuclear self-combustion, finally flared up. Yet now, star formation follows a different scenario in which it is not only giant stars that run the show. There is now plenty of scope for small, lightweight stars such as our own Sun.

☆ It is the gas molecules that make this possible. While a collapsing cloud of isolated hydrogen atoms becomes so hot that the gas pressure prevents further collapse – unless the cloud is colossally huge, giving rise to a gigantic monster star – molecules function as efficient radiators that keep the contracting cloud cool. Under the influence of the increasing pressure and temperature, molecules consisting of two hydrogen atoms begin to rotate, shake and vibrate. They then radiate forth the absorbed energy at long wavelengths – infrared and millimetre waves – that can escape the dark cloud unhindered.

☆ Because of this cosmic cooling mechanism, the temperature in the condensing cloud remains relatively low and gravity is able to triumph over the gas pressure in the form of small, dark globules. Without the help of molecules, which did not yet exist in the infant Universe, stars like the Sun would never have been born and the Milky Way would be populated with a handful of wasteful giant stars that burn up far too quickly to offer life any chance to evolve.

Dark globules condense into new stars, causing the surrounding area to glow and producing shockwaves in thin clouds of gas and dust.

☆ And yet it is the giant stars that are also the first to appear in the molecular clouds. In the dark depths of the gas and dust cloud, in a condensation that is perhaps the result of accidental turbulence, matter plunges into the gravitational abyss at increasing speed. Agitated by internal eddies and boosted by increasingly faster rotation, the warming cloud disintegrates into fragments, each of which collapses in turn. Sometimes four or five, sometimes ten or twenty – the predecessors of so many stars, each of them considerably more massive than the Sun, that switch on one by one as the nuclear fusion reactions in their hot cores begin.

☆ So it is that a glittering star cluster – a swarm of sparkling lights in the black emptiness of space – is formed in the dark, hidden vaults of the molecular cloud. Almost fully grown but still invisible to the outside world, the cluster is like an unborn baby that is on the point of leaving the warm safety of its mother's womb.

☆ The young, massive stars produce gigantic amounts of energy: heat radiation, visible light and also ultraviolet radiation – photons so rich in energy that they evaporate dust particles, smash molecules apart and ionise atoms. It is as if an exasperatingly slow explosion is taking place inside the molecular cloud, blasting clean the space around the cluster, sending shock waves through the tenuous matter, creating a giant cavity in which the newly formed stars are lord and master.

☆ Sometimes, when this cavity appears close to the surface of the molecular cloud, it can burst like a boil. Then the dark outside wall reveals a deep crater with frayed edges of dust, irradiated from the inside and illuminated by the young nebula that is still surrounded by thin strings and scraps of the volatile matter from which it was born. This hot, glowing web of gas is racked with violent shock waves that produce new condensations, new globules in which gravity does its work, new collapsing masses which, in turn, disintegrate into separate protostars.

Everything is in a state of constant flux and nothing lasts forever

☆ A baby boom of new suns spreads like wildfire. Sometimes the violence subsides again after a short time, the baby boom does not really get started, and in the molecular cloud, a brief fire erupts here and there. But elsewhere the baby boom takes on epidemic proportions and the star-forming process passes a critical point: a true population explosion follows that, in a short time, consumes almost all the molecular cloud. This leads to the formation of huge stellar clusters – tens or hundreds of thousands of cosmic power stations in which cool gas and dust from the original cloud are burned up and converted into radiation and new elements.

☆ All of this takes place at the leisurely rate of the Cosmos, in which every blinking of an eye lasts a century and metamorphoses occupy time-scales of millennia. The heartbeat of the Cosmos, the pulse of the Universe, is as slow as the rhythm of geology on our own home planet, where continents drift over the Earth's surface, plates collide or flow into the ductile mantle, chains of mountains are thrust upwards and where everywhere the staccato hammering and rattling of earthquakes and volcanoes reverberates. Just as much as our enduring impression of the Himalayas, the apparently timeless permanency of the Universe is an artefact of human perception, because everything in nature is in a state of constant flux and nothing lasts forever in a Cosmos in which microseconds, months and millennia are only arbitrary markings on the ruler of time.

☆ However, there is no trace of mankind as yet and the countless atoms that will one day make up Mount Everest are scattered over thousands of cubic light-years. And yet the comparison with the geology of the Earth is not so far-fetched: in the star-forming regions

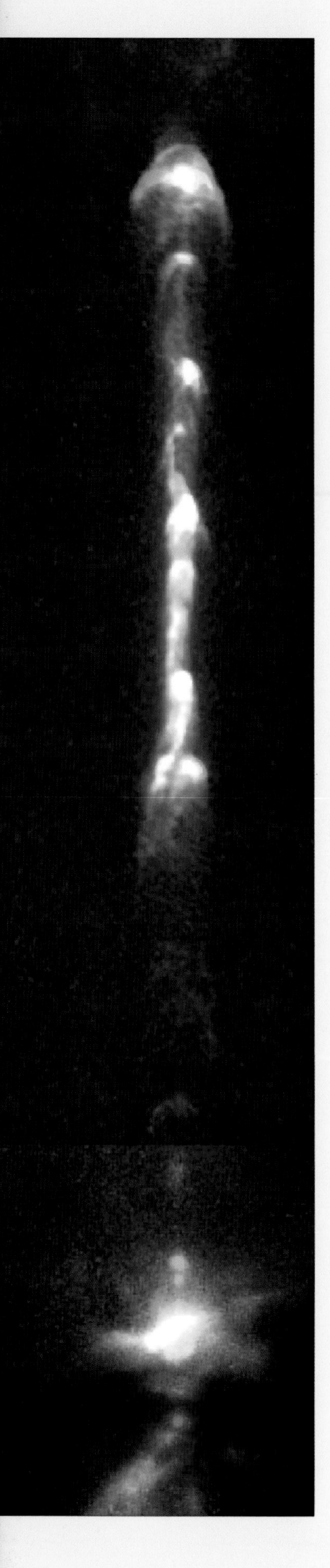

abrasion, erosion, moulding and landscaping are also taking place. If we were to speed up the film and reduce millennia to minutes, we would see how dark dust clouds are sandblasted with ultraviolet starlight, how matter is scraped away by energetic photons, just as canyons are dug by water erosion, how denser, more stubborn clouds remain, sculpted into fanciful trunks and pillars in a cosmic version of Monument Valley.

☆ On the tops and flanks of these dark stalagmites, the same photon sculpting process is taking place on a smaller scale and miniature versions of the dark dust columns are formed, sculpted by ultraviolet radiation like their much bigger counterparts and oriented towards this fiercely radiant star cluster. The interface between the cool, dark molecular cloud and the hot, expanding bubble of ionised energy changes into a fractal landscape of thin, glowing gas nebulae and dust condensations in which matter is once again compressed into small globules, hot protostars, young suns.

☆ And new stars are coming to life not just here in the immediate vicinity of this cluster but also at other places in the molecular cloud, in other cosmic delivery rooms. And not just in this cloud but simultaneously in the countless other molecular clouds that inhabit the spiral arms of the Milky Way. And not just in the Milky Way but also in millions of other galaxies. The wonder of birth is taking place all over the Universe, not just now but for billions of years already and for billions of years to come.

☆ Even though the newly born protostars hide behind impenetrable clouds of interstellar matter, even though their first light and heat is still concealed, they often give away their presence by the violent turmoil that they cause immediately around them, just as a newly born baby announces its arrival to the world far outside the delivery room with its piercing cries. Young protostars emit energetic X-rays – the electromagnetic equivalent of a shrill, high-frequency scream – and also spew forth jets of quickly-moving gas into space like miniature versions of the energetic jets of active galaxies and quasars. Not a lot is known about where they come from but here too the jets point in two opposite directions along the young star's axis of rotation. The jets break straight through the surrounding gas and dust clouds, pushing cool matter ahead of them until they are finally slowed down many light-years further on and come almost to a standstill. While the protostar itself is invisible, hidden in the dark mists from which it originated, it is flanked at a great distance by two hot clouds of swept-up matter, each with its own shock fronts and bow waves.

A young protostar blasts a beam of matter into space, causing shock waves in the surrounding gas.

When the last filaments of gas have been blown away in a large star-forming area, it reveals a sparkling star cluster of many hundreds of suns.

Newly born stars heat up their surrounding area, bathing it in the rosy light of glowing hydrogen gas.

The solar wind of a giant star, out of shot on the right, causes a bow wave around a smaller star which, itself, blasts matter into space.

☆ So what are these stars that first see life in the molecular clouds of galaxies? They are colossal, giant stars, tens of times more massive than the Sun, that blow huge amounts of powerful radiation into space, but also smaller, cooler stars that will perhaps one day shine their golden yellow light and gentle warmth on alien landscapes and germinating life-forms. These dwarf stars are less impressive and less luminous than the grotesque blue-white giants and supergiants but they occur in far greater numbers. These are the true "muggles" of the Milky Way.

☆ Even more numerous, and less impressive, are the red dwarfs. These are much smaller and cooler than the Sun – nondescript pips of weak, blurry light that cannot be seen with the naked eye at a distance of more than one light-year. The pressure and temperature in the cores of these stellar half-pints are so low that the nuclear fusion reactions do little more than simmer, using hardly any fuel. Yet because of this, they can continue to produce their small amounts of energy for many billions of years.

☆ And finally there are brown dwarfs that actually radiate a dull, magenta glow. Ten times smaller in diameter and fifteen times lighter than the Sun, these warm balls of gas are the size of the planet Jupiter and have insufficient mass to start the hydrogen-burning process at all. These are failed stars – stunted premature babies. And they are so difficult to observe that their true number is unknown.

☆ Thus, over billions of years, each galaxy produces an enormous variety of stars that vary not only in size, mass, luminosity and life expectancy but also in colour. The Cosmos is

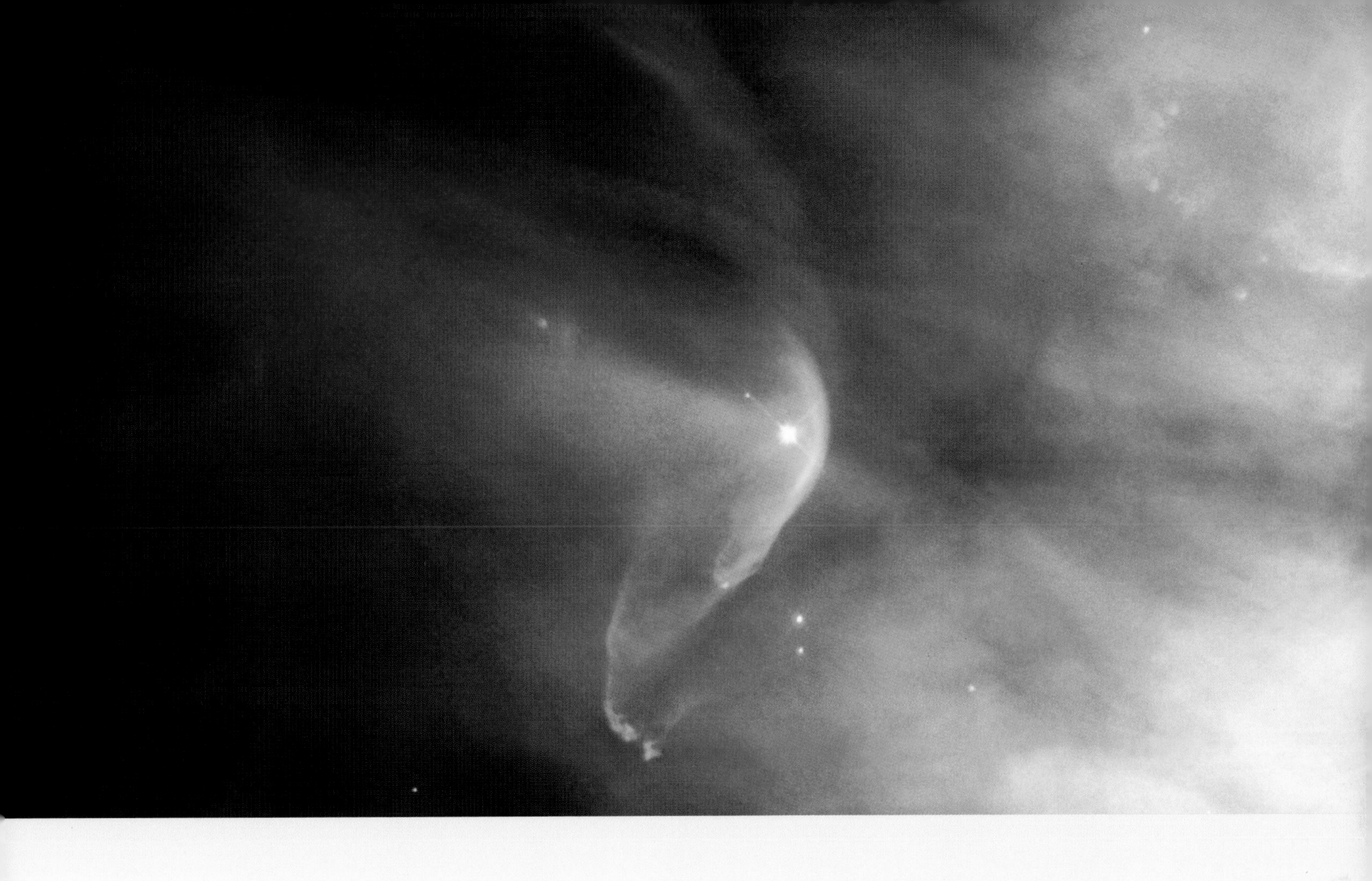

inhabited by a colourful array of stars, a many-hued population in which just about every shade in the spectrum is represented. The heaviest stars of all are the biggest and brightest with a surface temperature of tens of thousands of degrees. Because of this, they radiate a blinding blue light and are even brighter at the more energetic ultraviolet wavelengths. Slightly lighter stars are less hot and luminous and produce a white light. Stars like the Sun, with a surface temperature of six thousand degrees, are yellow, while the smallest and coolest of dwarf stars are orange or red.

☆ And in all of these stars, irrespective of their origin or colour, the same nuclear fusion process is taking place that earlier, a few minutes after the Big Bang, determined the original composition of the universe: the light nuclei of hydrogen atoms – in fact, individual protons – fuse together to form the heavier nuclei of helium, consisting of two protons and two neutrons. On this subatomic scale, the evolution of the Universe is no longer ruled by gravity or electromagnetism. It is now the strong and weak nuclear forces that determine how matter behaves. In fact, this transfer of power is possible only because of the conditions shaped by the two other forces: the nuclear forces have an extremely small range and can do their work only because the pressure inside the star (a direct result of gravity) is high enough to press positively charged protons together, in defiance of their mutual electromagnetic repulsion.

☆ The nuclear metamorphosis of hydrogen to helium – the first phase in the production of all chemical elements in the world around us – is a step-by-step process. Two protons collide. Owing to the ejection of a neutrino and a positron, one of the two changes into a neutron – a

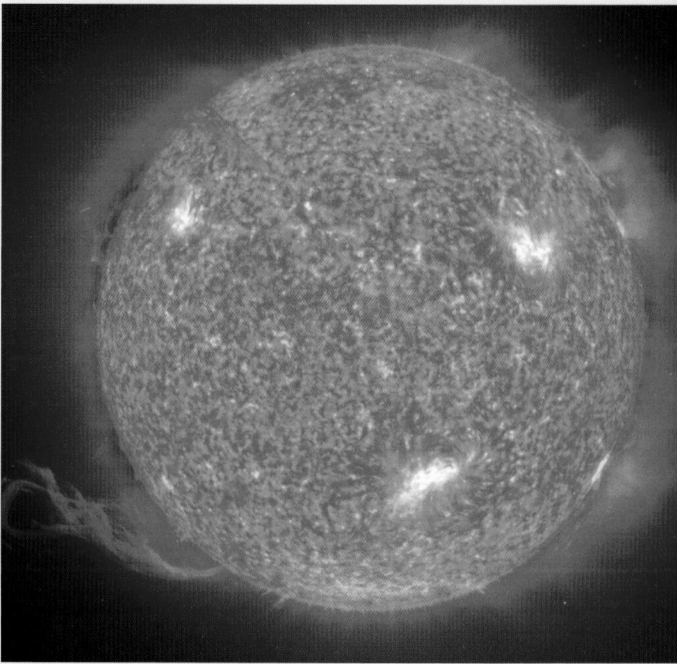

A newly born star still shrouds itself in the remains of the gas cloud from which it was created (*left*). Our own Sun is a middle-aged star (*right*).

nuclear particle without an electric charge. The strong nuclear force bonds this neutron to the second proton to form a so-called deuterium nucleus. This is a rare reaction: at the centre of a star like the Sun, one individual proton may go a billion years alone before fusing with another proton to create a deuterium nucleus. However, the number of protons at the centre of a star is so inconceivably huge that there is still an impressive production of deuterium – heavy hydrogen.

☆ A deuterium nucleus does not remain alone for long. Within a second, it has struck up a relationship with a third proton. This produces a nucleus of helium-3, a light variant of helium consisting of two protons and one neutron. The reaction also produces an energetic photon – a 'particle' of radiation. This is the nuclear reaction to which we owe the light and warmth of our Sun. After about one million years, the helium-3 nucleus collides with another helium-3 nucleus – a reaction in which four protons and two neutrons take part. The two neutrons and two of the four protons continue life as a very stable nucleus of helium-4. The two remaining protons go their way and can once again take part in the nuclear fusion process.

☆ In fact, in most stars, there is also another chain reaction taking place in which atoms of carbon, oxygen and nitrogen play a role. This CNO cycle, which in fact is dominant in stars more massive than the Sun, is more complex but the net result is the same: four hydrogen nuclei are converted into one helium nucleus. The fusion reactions themselves may be relatively rare, but the amount of hydrogen that is converted into helium every second in the centre of a star like the Sun is truly amazing: five hundred and seventy million metric tons. A

Pages 48/49: In cosmic delivery rooms, dark clouds of dust are illuminated by brilliant suns while hidden protostars drive shock waves through radiant gas nebulae.

relatively small part of this – less than one per cent, which is still four million metric tons per second – is converted directly into energy. Stars are gigantic fusion reactors with inconceivable capacity – hydrogen bombs before which the destructive capability of a human arsenal pales in comparison.

☆ The vast majority of helium atoms in the Universe were formed in the first couple of minutes after the Big Bang, when the entire Cosmos was as hot and compact as the centre of a star and fusion reactions set the Universe ablaze. But in this primordial, hectic period hardly any heavier elements were formed: the inferno did not last long enough and the Cosmos cooled down too quickly. The Big Bang alone was not able to produce the chemical diversity necessary for the formation of planets and moons, plants and people.

☆ But now stars of the most diverse masses, dimensions and luminosities shine everywhere. Now the chemical evolution of the Universe can continue from where it left off. No longer is the entire Cosmos one gigantic nuclear fusion power station that shuts down after a couple of minutes: now small nuclear ovens are operating in hundreds of billions of places in the Milky Way and in just as many stars in hundreds of billions of other galaxies. Admittedly, they are not fired up as high but they continue to do their work for hundreds of millions or even billions of years.

☆ The work of gravity in providing order, the hierarchical process of gravitational collapse and fragmentation, has transformed the seething ocean of the Big Bang into a huge emptiness in which matter is concentrated in variously shaped galaxies that, in turn, have produced countless radiant suns. Not only do these supply the indispensable of energy – light and heat – for every conceivable form of life in the Universe, but the constant fusion reactions in their hot centres are also essential to the production of the complex molecules from which all life is built.

We are, in the most literal sense, stardust

☆ Without the birth – and death – of stars, the formation of the Earth would have been impossible and the origin of mankind would never have happened. The birth of life cannot be regarded separately from the billions of years of evolution of the Cosmos, or from the recycling of matter. The atoms in your body are the products of cosmic fusion reactions. We are, in the most literal sense, stardust.

5

Recycling

Gigantic stellar explosions blast new elements into space. Carbon, oxygen, silver and gold – brewed inside stars and returned to the evolving Universe

LOOK AROUND YOU AND OBSERVE THE rich diversity of nature. Colour and movement, dead and living matter, quiet simplicity and boundless complexity – an infinite palette of objects and phenomena built from nothing more than a handful of basic building blocks and ruled by four forces of nature. From the first moments after creation, the Universe has developed from a practically formless mist of identical particles to a wondrous world of galaxies, protoplanets, sunflowers and sonnets. Hydrogen and helium have been there almost from the beginning but, other than that, everything in our field of vision owes its existence to the evolution of the Cosmos.

The lamp by which you read this book has a copper wire, a tungsten filament in a silicon-oxide housing, perhaps a base made of carbon chains or a decorative strip made of chrome – each of them elements that did not yet exist fourteen billion years ago. The iron in your blood, the calcium in your bones, and the phosphorus in your cerebral cortex are all products of the evolution of the Universe. Silver, gold and platinum are cosmic precious metals.

But it wasn't easy. Building new elements is a laborious process that can take place only under extreme conditions. The Big Bang did not last long enough; instead, the necessary matter had to be swept together into clusters and galaxies, gas clouds had to collapse and stars had to be built in which nuclear fusion reactions could do their work undisturbed for billions of years. Only in the stellar melting-pot can the chemical mix be brewed that lends the Cosmos its chequered appearance.

So the question remains of how the melting-pot was drained. The gradual, step-by-step production of heavy elements in the centre of a star is not the complete answer. Only when these elements are given back to the Cosmos, poured forth into a less extreme environment where further work can be done on forming fragile molecules and complex compounds, only then is the way clear for the creation of planets and life. And for that, stars must die, cosmic nuclear power stations must vaporise and explode – a process of dismantling that produces the most bizarre remnants.

However, chemical recycling in the Cosmos does not depend exclusively on death and destruction, on all-destroying supernovae and exploding stars. Stars also lose matter during their lives; they exhale atomic nuclei and blow a steady stream of electrically charged particles into space. This stellar wind, most of which consists of electrons and hydrogen nuclei, also contains dashes of heavier elements churned up from the centre of the star by convection and turbulence.

Red rings in a neighbouring galaxy mark the place where a star exploded in 1987.

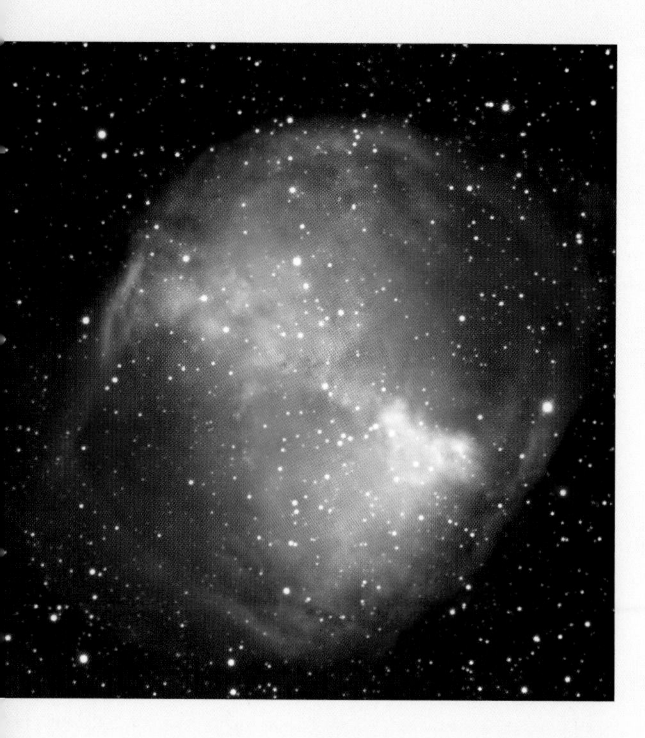

A star like the Sun blasts a thin film of gas into space at the end of its life.

The powerful particle radiation of a heavy, hot giant star causes wispy cirrus clouds of cosmic gas.

And at the later stages of evolution, when the star has expanded into a cool red giant, there is not only a greater loss of mass but the level of carbon, oxygen and nitrogen in this powerful stellar wind also increases.

But where does this spectacular growth, this impressive expansion of the star, come from? What causes this sudden change into a bloated giant? Once again, it is the nuclear reactions in its centre that are responsible. Quite shortly after the creation of a star, when the fusion of hydrogen to helium is well under way, the star's size, surface temperature and brightness are determined entirely by the delicate balance between gravity, which wants to compact the stellar gas further, and the pressure of the hot gas inside the star that resists this. But as soon as other nuclear reactions gain the upper hand, changing the internal pressure and temperature, the star seeks a new balance, which involves a new diameter, a new luminosity and a new surface temperature. In this way, the outward appearance of a star betrays its inner structure. Every sudden expansion or contraction directly mirrors upheavals in its hidden interior.

No star escapes these internal revolutions. If the hydrogen fusion lasts long enough, it creates an increasingly larger reservoir of helium and, sooner or later, this helium core will be massive enough to ignite spontaneously. While hydrogen combustion is still taking place in a thick layer around the core, helium is also now being converted into carbon inside the star. This produces so much energy that the star soon swells up like a cheese soufflé. Its surface temperature drops because the energy from its centre is spread over a much larger area. The result is a giant star with spectacular energy production and luminosity but with a cool, thin outer mantle that is barely held together by gravity.

Later, when carbon nuclei also fuse to form even heavier oxygen atoms, the process repeats itself and the star can grow even bigger. Little remains of its original, apparently permanent stability: the star swells and shrinks, heaves and puffs, sweats and blows, and soon large quantities of cool gas are blown into space that quickly form into fragile molecules, small crystals and particles of smoke and soot. In a cosmic blinking of an eye taking a few tens of thousands of years, these exhaled mists light up in the most fantastic colours, irradiated by the surviving star, which is now smaller, whiter and hotter. After this, the clouds are blown away forever into the dark space between the stars where, eventually, they will be part of a new star, a new source of light and heat, a new phase in cosmic recycling.

In the case of stars that are born as heavyweights, the life-cycle is different. And faster, much faster. Because of the high temperature and pressure at their centres, the speed of the fusion reactions is also much higher. The nuclear fuel is burned up at a terrific rate; energy production

is many thousands of times greater than that of a star like our Sun and when the giant star starts burning carbon, after no more than a few tens of millions of years, not only is oxygen created but also heavier elements such as neon, sodium and magnesium. And always, when a particular fusion reaction comes to a stop through lack of fuel, the star contracts under its own weight until the core temperature is high enough to start the next reaction, again creating heavier elements such as aluminium, silicon, phosphorus and sulphur.

So, the star ends up like a cosmic Russian doll in which every nuclear furnace is surrounded by a slightly cooler layer with a different, less energy-rich fusion reaction. Beneath the star's unstable outer layers, which are blown into the Universe at high speed, there is still a thick shell in which hydrogen is converted into helium. Inside this is a layer in which helium nuclei are fused into carbon. And inside this is a layer in which carbon is fused to make oxygen. Then a layer of oxygen burning, then one of silicon burning. And, at the very heart of the star, is a steadily growing core of iron atoms, each with twenty-six protons and thirty neutrons – fifty-six times heavier than the nucleus of a hydrogen atom.

Iron is not a precious metal. Neither is it an indispensable ingredient of life. It plays no crucial role in our everyday world. But a very special place has been reserved for iron in the periodic table of the elements, in the order of nature. When light atomic nuclei fuse to make heavier nuclei, energy is released, from which emanates the life-bringing radiation of the Sun. Energy is also produced when heavy atomic nuclei are split into lighter nuclei, which was demonstrated in a horrible way in 1945 with the bombs that were dropped on Hiroshima and Nagasaki. But iron is different. To split or fuse iron atoms, you have to *add* energy. With the formation of iron atoms in the core of a massive star, the process of spontaneous nuclear fusion comes to a stop.

For millions of years, the star has resisted gravity. The production of energy in its centre has always provided sufficient counter-pressure to prevent the further collapse of the star. A new balance has always been found for every new nuclear reaction – a new balance between

The Cosmos becomes enriched with all of the elements needed for the formation of the Earth

gravitational and nuclear forces. Every star in the sky is the scene of a battle between the forces of nature – a nuclear battle for life and death. And now that one of the warring parties has given up the fight, the other is automatically victorious.

Gravity seizes its opportunity. What started long ago with the collapse of an interstellar cloud of gas and dust can finally be completed. No longer impeded by opposing forces from the hot centre of the resulting star, gravity draws the matter closer and closer together. The core is compressed by the unbearable weight of the hundreds of thousands of kilometres of the star's thick outer layers. Now that gas and radiation pressure inside the star have given up the ghost, the outer layers collapse into a fatal free-fall, just as a hang-glider drops when the thermal suddenly ends.

Is gravity now finally all-powerful? Will all matter soon disappear into a bottomless black hole, torn from the familiar environment of space and time as we know it? No. Deep inside the collapsed giant star, feverish preparations are being made for a new phase in the final battle. At subatomic level, matter conspires to oppose gravity one last time. In a fraction of a second, the star's iron core collapses. From a diameter of ten thousand kilometres to a thousand, then to a hundred and finally to no more than thirty kilometres. Three hundred times smaller in diameter; more than twenty-five million times smaller in volume. Not even the stable iron atoms are a match for such violence. They are broken up into separate protons and neutrons that swarm around at lightning speed in a seething sea of free electrons.

However, this collapsed stellar core is so compact that all of these subatomic particles are literally pressed onto and into each other. Negatively charged electrons penetrate the positively charged protons and sacrifice themselves to a quantum process in which one of the up-quarks in the proton changes into a down-quark, creating an uncharged neutrino. This invasion of electrons neutralises the protons into new neutrons and, in a short time, the small core of the star changes into an immensely hard ball of stiff, compressed neutrons – a couple of times more massive than the Sun but only a few tens of kilometres across.

In the immediate vicinity of one of the heaviest stars in the Milky Way, bright gas nebulae are stirred up and dark clouds of dust are pressed together.

The formation of the neutron core – not at the dawdling speed that we have become used to in the Cosmos but literally in a fraction of a second – saves the star from its final decline. And even more importantly, thanks to this quantum reaction, the star's collapse becomes an explosion. Heavy atoms are thrown into interstellar space and the Cosmos becomes enriched with all the elements needed for the formation of the Earth, living organisms, and you and me.

If you were to see a massive giant star collapse at the end of its life because nuclear reactions are no longer taking place inside it, you would probably find it difficult to imagine how nature can succeed in stopping this free-fall and turn it into a titanic supernova explosion. Many trillions of tonnes of stellar gas collapse inwards at unbelievable speed; the star shrinks visibly and it looks as if it will vanish from the stage in a few moments. In this angry maelstrom, pressure and temperature rise to unprecedented values and immediately the fire of cosmic alchemy ignites. Never before in the life of the star have the fusion reactions been so violent or new elements made at such a terrific rate.

But this nuclear inferno is brief and is incapable of halting the collapse of the star.

A supernova in a distant, dust spiral nebula radiates almost as much light as the millions of stars that together make up the bright core of the galaxy.

Shimmering matter tumbles downwards burning and screeching; atoms are ripped apart as quickly as they are produced. The star looks almost like a reversed miniature version of the Big Bang. But deep at the heart of this dying giant, like the incompressible pip in a crushed fruit, lies the neutron core, as rigid and solid as a thirty-kilometre atomic nucleus – an impregnable bastion with a density of a hundred billion kilograms per cubic centimetre. This iron ball – its crust is actually composed of compressed iron atoms – is the bottom of the gravitational well, and when the outer layers of the star dash themselves against it at speeds of many tens of thousands of kilometres per second, a gigantic outward-going shock wave is created. Helped by unpredictable, turbulent eddies and currents in the stellar gas, and by the energy of the neutrino tidal wave created when the neutron core was formed, the shock wave succeeds in halting the catastrophic collapse of the star just in time.

★ Now the chaos is complete: while the outer layers of the star are still falling inwards, a ball of fire rushes out of the depths which stops and reverses the collapse like a spherical snowplough, turning the merciless implosion into an unstoppable explosion. Heavy and light atomic nuclei, loose protons and neutrons, electrons and neutrinos all play their part in the cacophony of nuclear reactions emanating from a star turned inside out. In this way, in the short death-agony of the star, all the elements and isotopes in the chemical repertoire are created, from cobalt, nickel, copper and zinc to silver, platinum, gold and mercury; from stable

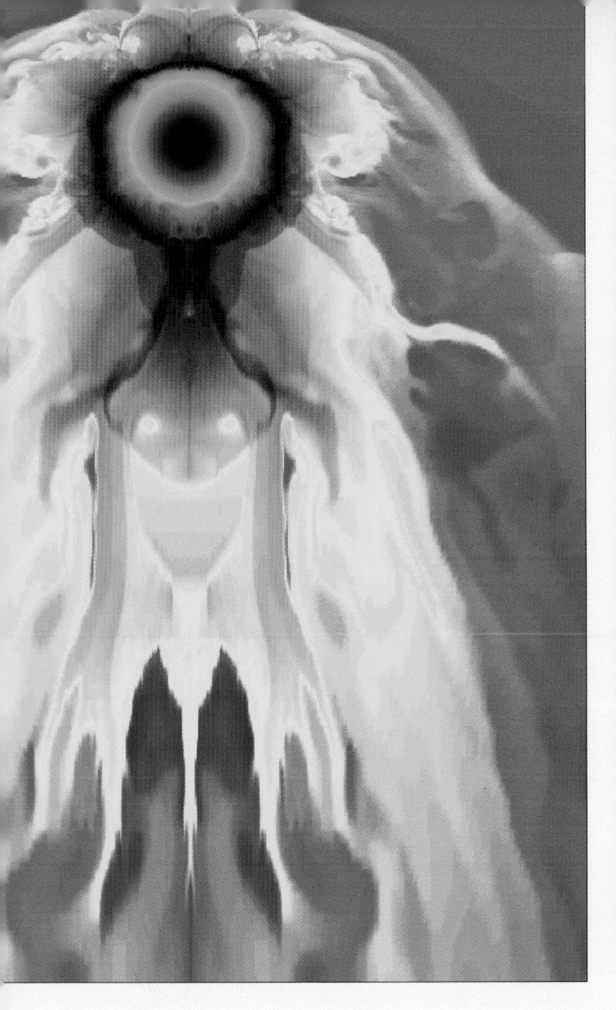

Supernovae cause shock waves, here in a computer simulation (*left*), and leave behind a rotating pulsar with revolving beams of radiation (*right*).

elements like xenon, caesium and tungsten to radioactive substances such as radium, uranium and plutonium.

☆ Glowing from the radioactive decay of unstable isotopes, the explosive cloud expands and, for a few weeks, a 'new star' can be seen in the sky from up to millions of light-years distant, brighter than all the stars surrounding it. Here and there, the speeding hot matter is slowed down slightly by shreds of gas blown away by the star at an earlier stage and there the glowing, spherical supernova remnant quickly changes into an irregular nebula that will get larger, thinner and cooler over the years, decades and centuries. Finally, the cloud is blown away and vanishes, and the star's ashes are scattered in the void – a new addition to the cosmic ingredients from which a new generation of stars can be created.

☆ Yet look: not the whole star has exploded. The small, compact neutron core remains – the super-condensed mortal remains of a once so imposing giant star. This bizarre celestial body turns on its axis tens or even hundreds of times a second like a mad humming-top, thanks to the law of the conservation of angular momentum – the same law of nature that makes a figure-skater turn faster as she pulls in her arms during a pirouette. Because of the contraction, not only has the speed of rotation increased enormously but also the strength of the magnetic field. Electrically charged particles from the surrounding area are directed to the magnetic poles where they strike the surface at great speed and create hot spots that transmit powerful beams of radio

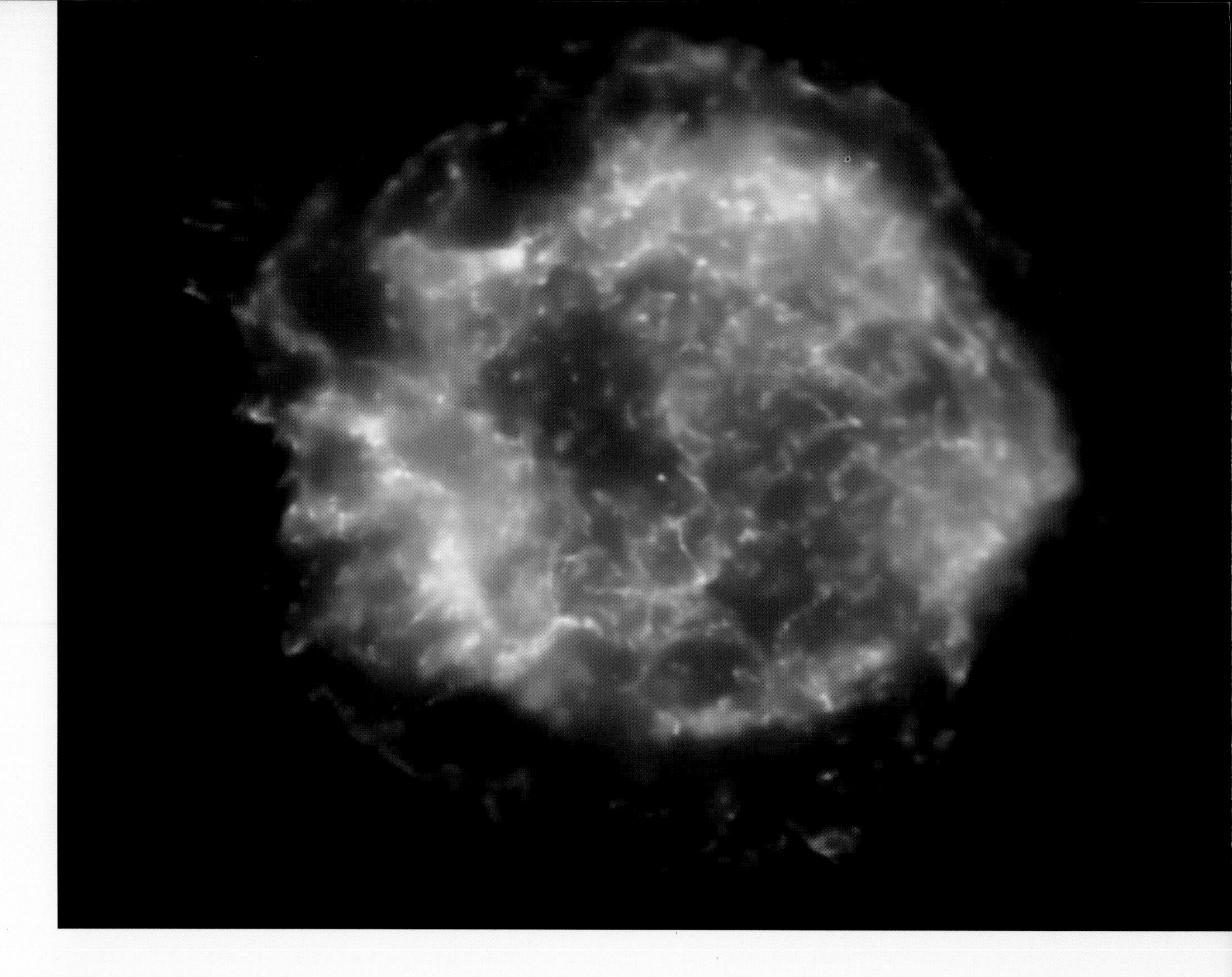

Young supernova remains are so hot that they transmit powerful X-ray radiation

The remains of this supernova are not yet a thousand years old – the ejected matter from a star that exploded in the year 1054.

and X-ray radiation into space. Tiny quakes of the neutron star's iron crust lead to disruptions in the magnetic field, which, in turn, can result in sudden eruptions of powerful X-ray and gamma radiation.

Countless neutron stars race like cosmic zombies almost invisibly through the Milky Way galaxy, often at high speed, when the supernova explosion that produced them was rather asymmetrical, so that a net force in one direction remained. Yet sometimes we can actually see neutron stars. Because the hot magnetic poles do not normally coincide with their axis of rotation, the beams of radio radiation produced there sweep through the Universe like the light-beams of lighthouses. If you were to find yourself in the path of one of these beams you would see the neutron star flash on and off many times per second. So unchanging is the period of rotation of such a pulsar that these cosmic strobe-lights are some of the most accurate clocks in nature.

When the supernova eruption takes place in a binary system, the remaining neutron star will eventually be visible from afar. When the smoke of the explosion has dissipated and the foggy residue of the supernova has blown away, initially a remarkable couple remains behind, engaged

in a hectic gravitational waltz. But when the ordinary star has evolved into a red giant, the outermost gas layers will be sucked away by the strong gravitational field of its compact partner. The gas accumulates in a rapidly rotating disk and then spirals irrevocably to the surface of the neutron star. The temperature increases; the accretion disk and the hot spots on the surface transmit powerful X-ray radiation. Such an X-ray binary shouts its presence from the rooftops to every corner of the Milky Way.

With its jets of radiation, magnetic eruptions, plasma clouds and accretion disks, the world of neutron stars and pulsars is a world of extremes, of demented speeds and temperatures, of immense explosions and crackling particles. There is no life-form that could survive in this environment of bending, snapping magnetic field lines, of all-penetrating cosmic radiation and scorching energy. These are the waste products of cosmic evolution, the slag and cinders of the celestial power stations that have brought forth the atomic building blocks of life. These are the pathological celestial bodies, good for nothing, but which are, at the same time, the necessary remains of the universal recycling of which we are a part.

And it can get even crazier than that. An X-ray binary can even vanish entirely in a

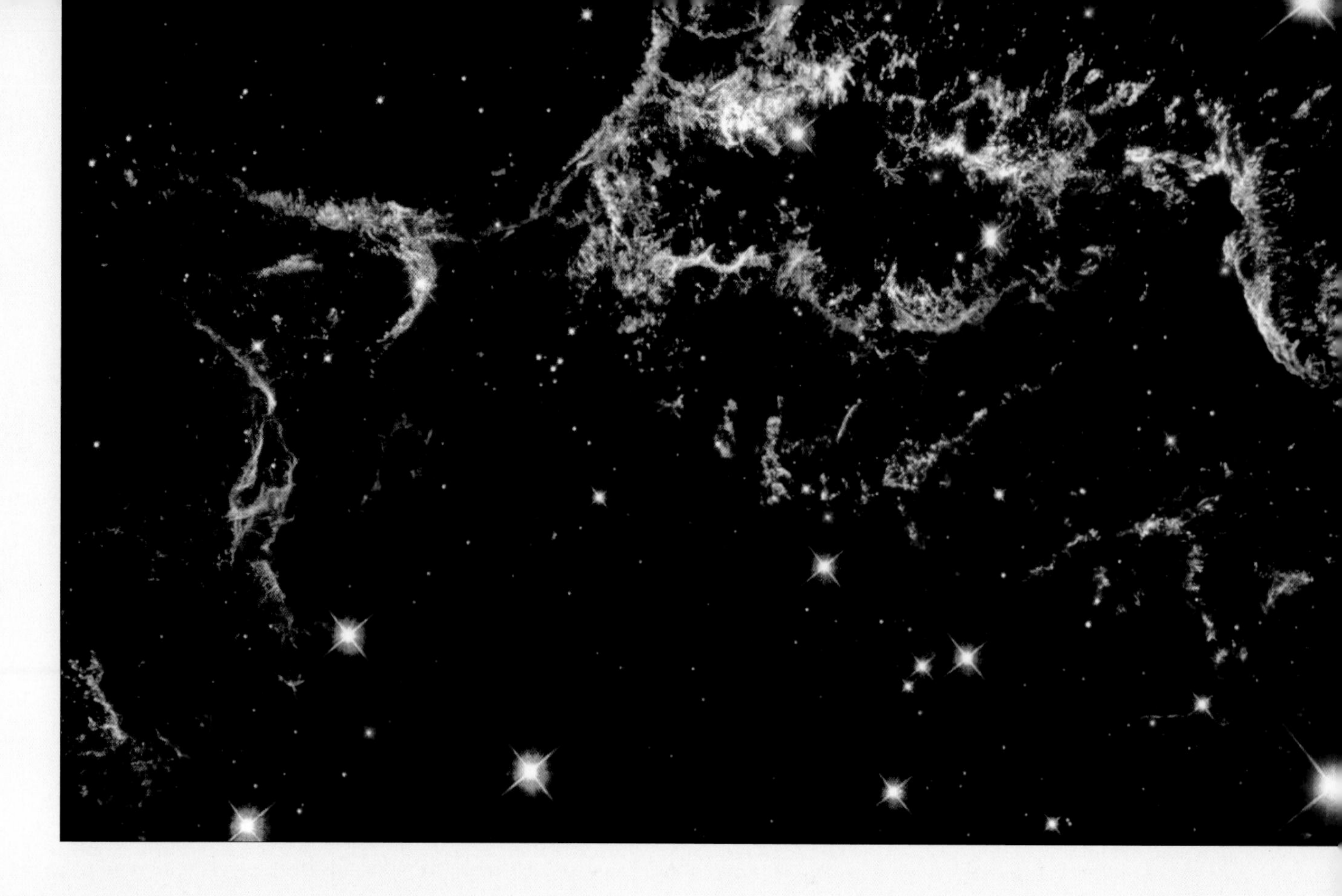

These cooled gas filaments are part of the remains of the youngest supernova in our own Milky Way.

catastrophic explosion. If the neutron star's dancing partner is massive enough and also goes supernova itself, this leaves a binary neutron star – two revolving celestial corpses moving around each other at terrific speed. The acceleration resulting from this is so gigantic that it causes disruptions in the basic fabric of space-time: gravity waves that move through the Universe at the speed of light and slowly but surely extract energy from the system. The results can be guessed: the two neutron stars revolve faster and faster, closer and closer to each other, until they are finally torn apart by their mutual gravitational pull and merge into an all-devouring black hole – a bottomless well of gravity that even holds light fast in its grip.

And if the original star – the massive giant star with which it all started – has sufficient mass and rotates quickly enough, the black hole can be formed immediately, during the supernova explosion. Then even the compact ball of compressed neutrons is not strong enough to bear the colossal pressure of the collapsing star and its matter will be compressed even more strongly, past the point at which the other forces of nature can offer resistance. Then gravity really does have mastery and a large part of the star vanishes forever through the one-way door of a black hole, beyond the horizon of gravity. One last death-cry of the dying giant still echoes for millions of years through the Cosmos in the form of a huge burst of gamma rays, the most energetic form of radiation in nature, and, for a very brief moment, this gamma-ray burst shines brighter than millions of galaxies.

The death of a massive star is a true hell – an inferno in which matter and energy, particles and forces are tested to destruction and where the laws of nature no longer appear to apply.

Pages 62/63: A star like the Sun will never explode but constantly blasts hot gas into space.

Neutron stars, pulsars and black holes are the frightening manifestations of this devilish process, the grim remains of this macabre death struggle. But while matter is being ripped apart by uncontrolled energy, particles and radiation are being sucked into a black hole and space-time is being deformed and twisted by gravity, elsewhere in the Universe, in a quiet, dark corner of the Milky Way galaxy, a start is being made on the formation of a new star, a new sun. The expanding supernova remnant, the nebula formed by the same explosion that created a neutron star or a black hole, gently collides with a tenuous cloud of interstellar matter, a cool whisp of gas and dust. Slow shock waves cause local density enhancements and, soon after, part of the cloud, sprinkled with the heavy elements created in the supernova, collapses under its own weight, very slowly at first, but then faster and faster.

The death of a massive star is a true hell

★ The birth of the Sun begins. A little while longer and there will be a new planetary system here. A short time later and the miracle of life will take place. And some of the atoms in this small, dark globule will one day be part of a human being, a writer, a reader. In just a little while.

6

Accretion

Cold planets are stuck together from dust, grit and pebbles – the by-products of the birth of a new star. There are gas giants and lumps of rock, but also a blue speck: the Earth

PLANETS PLAY NO SIGNIFICANT ROLE IN THE UNIVERSAL PLAN. Galaxies are the awe-inspiring building blocks of cosmic architecture; stars are the sparkling alchemists that lend the Universe its chemical diversity, but planets are no more than waste crumbs – coagulated refuse left over from the birth of a star. So, it's a strange irony of nature that the most complex phenomena in the Universe – DNA molecules, micro-organisms and consciousness – can take shape and flourish only on these insignificant heaps of matter, on the grey grit accumulated in the seams and cracks of cosmic evolution.

✧ Nothing in nature is perfect. Chance and chaos are the antagonists of control and order. Every rule has its exceptions and no single process is absolute. Even the fundamental law of nature describing the cosmic striving towards a constantly increasing degree of disorder – the second law of thermodynamics – apparently refuses to operate in certain circumstances. For example, an amorphous cloud of matter can suddenly reveal complicated patterns and structures. It is precisely because of this unpredictable, chaotic behaviour of nature that the present Cosmos is not a sterile world of stars that simply shine in an absolute vacuum but that thin curtains of gas and dust also drift through space, that stars are also accompanied by a complete retinue of planets, moons, asteroids and comets, and that life has obtained a foothold on at least one of these cold pebbles.

✧ A planetary system is an almost inevitable by-product of the formation of a star, at least when the cloud from which the star was created also contains sufficient heavy elements. The cloud collapses under the effect of gravity, at first almost unnoticeably and then faster and faster. In addition to this, there are forces at work that prevent all matter from finally ending up in the star. The gravitational collapse is not perfect, not one hundred per cent symmetrical, for the simple reason that the original cloud was not perfect or symmetrical either. Every molecule, every insignificant particle of dust moves in its own way and these individual movements do not average out exactly. The cloud as a whole has a slight net rotation – the vector sum of all these minuscule, arbitrary contributions. And rotation is, by definition, asymmetrical. It implies a preferred direction – an axis of rotation.

In the Orion Nebula, the growth of planetary systems is hampered by the powerful radiation of the four bright stars in its centre.

✧ In the case of a tenuous cloud a couple of light-years in size, this is not noticeable: the cloud doesn't even realise it itself. But the smaller it is, the faster it starts to rotate. The total quantity of rotational energy – rotational angular momentum in the jargon of physics – stays the same,

Dusty protoplanetary disks around newly born stars form the birthplaces of other planetary systems.

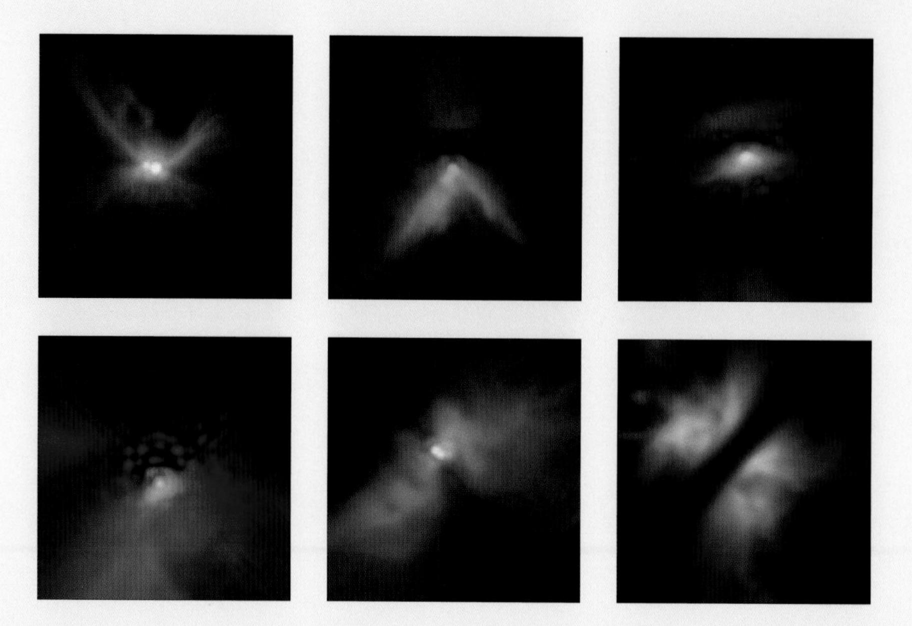

and a decrease in diameter is inevitably accompanied by an increase in the speed of rotation. This affects the shape of the contracting cloud. Even if it had been perfectly spherical originally – which, of course, is totally inconceivable in an imperfect universe – it would still be completely flattened by this fast rotation, like a lump of dough thrown into the air by a pizza baker, to fall as a flat pancake.

✧ Here is the order that automatically emanates from apparent chaos: whether the cloud is turbulent or static, fancifully shaped or very symmetrical, the process of collapse always results in a compact, spherical protostar, surrounded by a rapidly rotating flat disk of gas and dust. It is in this circumstellar disk that, in a few million years, planets will appear.

✧ At least, this will happen if the disk is left to its own devices for long enough. If it is not blown away by powerful ultraviolet radiation from neighbouring giant stars soon after its creation. If it does not evaporate away before planets begin to form. For planets do not get a chance everywhere in the Universe; not everywhere is it quiet enough for the step-by-step construction of new worlds. In the immediate vicinity of young star clusters, in the floodlights and smoke of dazzling heavyweights, the protoplanetary disk of a newly formed dwarf star does not have a long life. The result is a planetary miscarriage.

✧ In the quiet areas of the Universe where small, isolated groups of stars are created, far from the violence of the massive cosmic breeding machines, the precursors of new planetary systems get every chance. Naturally, the radiation from the mother star also has an erosive effect on the embryonic disk but it takes some time before the new star actually begins to shine and, also, it's not so easy to blow away a disk of gas and dust when you're at its centre.

✧ However, the lighter gas atoms and molecules, in particular, *are* blown away; larger, heavier particles are more difficult to eliminate. In the inner areas of the disk, close to the protostar where the temperature is high, almost all substances are vaporised and the cloud consists

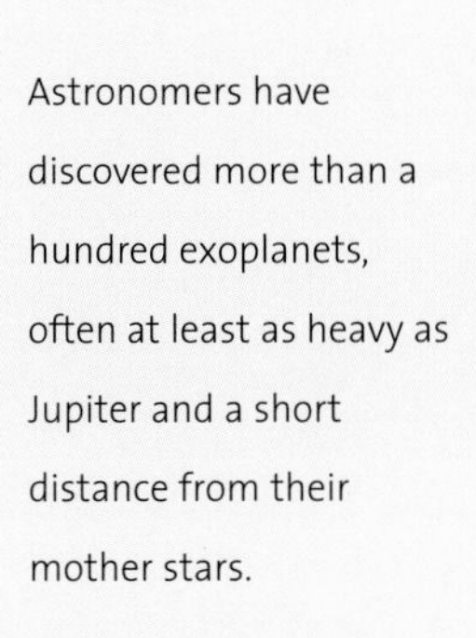

Astronomers have discovered more than a hundred exoplanets, often at least as heavy as Jupiter and a short distance from their mother stars.

almost entirely of gas. Here, only refractory matter with a high vaporisation temperature can survive, such as silicates and metals – elements and compounds that also occur in solid form under these conditions. These are the particles of dust and crystals that will later form 'Earth-like' planets – small worlds of metal and rock.

✧ At a greater distance from the star, the temperature in the protoplanetary disk is lower. Here, near the 'snowline', it is much too cold for volatile substances such as hydrogen gas and oxygen – light atoms that can easily be blown away by the young star. The atoms coalesce into water molecules that, in turn, stick together in small ice-crystals. The same applies to other compounds: carbon monoxide, carbon dioxide, methane, ammonia – all of these gases have their own 'snowline' and past this thermal demarcation line they condense into enormous

In the quiet areas of the Universe, the precursors of new planetary systems get every chance

quantities of relatively heavy ice-crystals that are not so easy to blow away. It is largely from these substances that the giant planets will later be created.

✧ While the star is still in its prenatal stage, while the centre of the collapsing gas cloud is becoming steadily denser and hotter but nuclear fusion reactions have not yet started, the first steps towards the formation of planets are taken in the protoplanetary disk. In human terms this process takes an eternity but in cosmic terms it happens at great speed: within a couple of million years, large celestial bodies are formed by the coalescing of this cloud of gas molecules, dust particles and ice-crystals, and the foundations of a complete planetary system are laid.

Swirls of cosmic dust, condensed from the waste products of stars, contain the building materials for planets like the Earth.

✧ First it is the slight molecular and electrostatic forces that make microscopically small particles of dust stick to each other to become small, fluffy, porous particles, which in turn merge to become grit and pebbles. In the outer parts of the cloud, this does not happen only with the relatively heavy dust particles but also with countless ice-crystals that mainly consist of much lighter elements. All of these fragments, pebbles and snowflakes – a millimetre or a centimetre in size – describe their own orbits around the star-to-be, each of them obeying the laws of gravity. But they also constantly push and rub against each other, creating larger structures. In time, it is not just molecular forces that are at work but also very slight mutual gravity.

✧ Once these heaps of matter have become big enough for gravity to start playing a role, this process of coagulation accelerates. Contrary to molecular bonding, which can happen only when two particles accidentally come very close to each other, gravity works also at a distance. Slowly but surely, the disk coalesces. Swarms of stones and ice-fragments are gradually drawn together into larger objects measuring a metre, ten metres, a hundred metres in diameter. There is no longer a way back: the planetary system starts here.

✧ The trillions of small fragments racing around the star-to-be like as many mini-planets – too big to be called lumps of ice or heaps of stone but still far from being fully grown planets – are called planetesimals. Their dimensions vary from a couple of hundred metres to a few tens of kilometres and they are the building materials for the new planets.

✧ Planetesimals are not numerous everywhere in the disk and they do not have the same composition everywhere. In the hot inner parts of the disk, where all the light gases have been blown away at an earlier stage, there are 'heavy' planetesimals formed from rock and metal. Their number is comparatively small because the primordial cloud from which the star was

The Sun was created from a contracting cloud of gas and dust.

In the rotating outer areas, matter coalesced into planets.

formed contained a relatively small amount of heavy elements. In the cooler outer regions, near the 'snowline', 'light' planetesimals are created, which consist largely of ice. There are huge numbers of these: firstly, because most of the protoplanetary disk lies beyond the 'snowline' and, secondly, because the lighter elements have not been blown away here but condensed into ice-crystals.

✧ This difference in number and composition of the planetesimals is crucial to the construction of the resulting planetary system. Close to the star, planets consisting mostly of rock and metal are formed – so-called 'Earth-like' planets. These are small for the simple reason that there are not so many rocky planetesimals around. In the outer regions of the disk, giant planets are formed that mostly consist of light elements. These gas giants are much bigger because the number of frozen planetesimals is simply much bigger.

✧ The accretion process continues and increases in intensity. No longer do small particles coagulate loosely together with each other; no longer is there a gradual gathering of pebbles and snowballs. Here, two planetesimals collide and are both pulverised by the

The big get bigger; the small go to the wall

energy released; there, they graze each other, are slowed down, heated and deformed, to merge together into a larger object with an even stronger gravitational field. Small fragments constantly smash into the surface of such an object and, from time to time, there is a large impact from a fully-grown planetesimal. The future planetary system gradually changes from a coagulated mixture of gas and dust into a hot battlefield of cosmic collisions, while the omnipresent gravity brings about the ultimate split in the population: the big get bigger; the small go to the wall.

✧ After a while, tens of hot protoplanets circle around in the inner parts of the solar system. These are thousands of kilometres in size – swollen by their gravitational greed and heated by the energy released by millions of large and small impacts. They are slowly sloshing balls of magma and molten metal enclosed in a thin, solidified membrane of rock that is still regularly punctured by the continuous bombardment of the remaining planetesimals.

✧ In the spectacular finale of the accretion process, the protoplanets also collide. Chaos and chance determine how many of these Earth-like planets will finally survive, how big they will be and in what orbit they will circle their mother star, at safe mutual distances and no longer menaced by monster projectiles. Still spluttering from the waning violence of the primordial bombardment, they finally come to rest, the metal core cools down and the mantle starts to solidify. Perhaps, one day, water will flow on one of these planets, a chain of mountains will arise, and a cell will divide – small miracles on a cool world, drowsing in the warm light of a young star.

✧ At the same time, a similar process of creation is taking place in the outer areas of the future planetary system. Icy planetesimals ram each other, burst into fragments or merge together; large objects clean their surroundings like gravitational vacuum cleaners; small, primitive celestial bodies melt from the impact energy of countless projectiles. Possibly, the countless hot protoplanets created here have small cores of metal and rock, a mantle of water many hundreds of kilometres thick and a brittle crust of ice. Here too, increasingly larger objects are being formed. Their immense gravitational pull also attracts the gas that has not yet been blown away from the protoplanetary disk. Thus, they grow into gigantic balls of gas with only a relatively small core of heavier elements. They are cooling gas giants with winds and eddies in their outer layers and a core where the pressure is so high that hydrogen gas is compressed into a liquid which, at great depths, even displays the electrical properties of a metal.

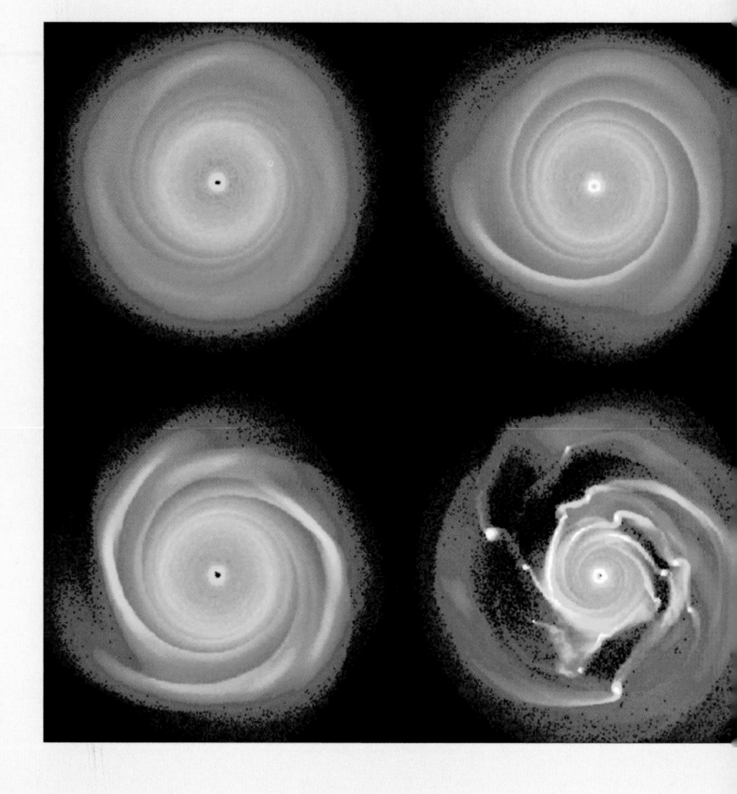

✧ Elsewhere in the Universe, where protostars may be surrounded by a much more massive disk of gas and dust, giant planets form quicker, not by the step-by-step accretion of ice-crystals, snowballs and planetesimals but by the stampeding gravitational collapse of an accidental condensation in the protoplanetary disk that then sucks gas, dust, grit, ice and larger objects inwards. The genesis is different but the final result is the same: irrespective of the properties of the protoplanetary disk, a handful of giant planets are created in the outer regions of the flattened, rotating cloud – colossal worlds of gas and liquid that seem to have little in common with the small Earth-like planets of rocks and metals closer to the star.

✧ But nothing in nature is perfect and the accretion process in a protoplanetary disk is no exception. Many planetesimals manage to escape collisions and avoid the gravitational suction of planets-to-be. They end up as rocky asteroids towards the centre of the system; as icy comets towards the outside. There are countless small scraps and fragments no more than a few tens of metres in diameter but also larger objects measuring a couple of hundred kilometres. While the planets get bigger and more massive and, slowly but surely, take on their final form, the remains of this growth process swarm around everywhere.

✧ This is cosmic driftwood, tossed back and forth on the waves of gravity: in the grip of the mother star but also subject to the gravitational field of the newly born planets. Some asteroids – the rocky planetesimals from the inner part of the system – inhabit a stable zone where they continue to orbit like miniature planets for many a long day. This belt of rubble is only occasionally startled from its sleep by a fresh collision, which produces yet more small fragments and cosmic dust. Other asteroids move in elongated, oblique orbits and constantly run the risk of colliding with a planet.

Computer simulations show that giant planets can be created rapidly by the fragmentation of a rotating protoplanetary disk.

In the early years of the Solar System, small planetesimals were swept up by the gravity of fully grown planets.

✧ Comets – the icy planetesimals at a greater distance from the star – are thrown this way and that by the gravitational perturbations of the giant planets, sometimes inwards where they collide with a planet or vanish forever into the inferno of the star, and sometimes outwards where they finally arrive in a huge cloud of deep-frozen ice at the cold, dark periphery of the planetary system.

✧ And it is not only small asteroids and comets that are ricocheted through space by the gravity of the giant planets like the steel balls in a pinball machine: neither are the Earth-like planets in the inner regions of the system sure of their lives. The rule of gravity is capricious and

Diversity in the Solar System: Jupiter (*left*) is composed mostly of gas, Quaor (*centre*) of ice and the Moon (*right*) of rock.

ruthless. Under the influence of its neighbours or the remains of the gas cloud from which it was created, a giant planet can slowly but surely spiral inwards and this planetary migration produces chaos in the original order. In this way, a beautifully symmetrical protoplanetary disk can finally result in a bizarre planetary system with gas giants at improbably small distances from the mother star, or in extremely elongated orbits. In such a crazy system, small, Earth-like planets find it impossible to survive. Long before the promise of life is fulfilled, the planet dies a premature death: ejected out into the depths of the Universe or swallowed up by the star that should have served as the guardian of life.

✧ Chaos, chance and imperfection – these are the ingredients from which planetary systems are made. Every star has countless identical twin sisters in the Milky Way alone: stars with the same mass, the same luminosity, the same composition. Protoplanetary disks display slightly more variation but planetary systems 'take the cake'. However similar the initial situation may be, the genesis of the system depends on such an unbelievable diversity of accidental factors that the result is always unique and unpredictable. How many planets are created, with what masses and dimensions, what orbits they describe around their mother star and whether migration occurs or not – the number of possibilities and combinations is endless and no two planetary systems in the Universe are identical.

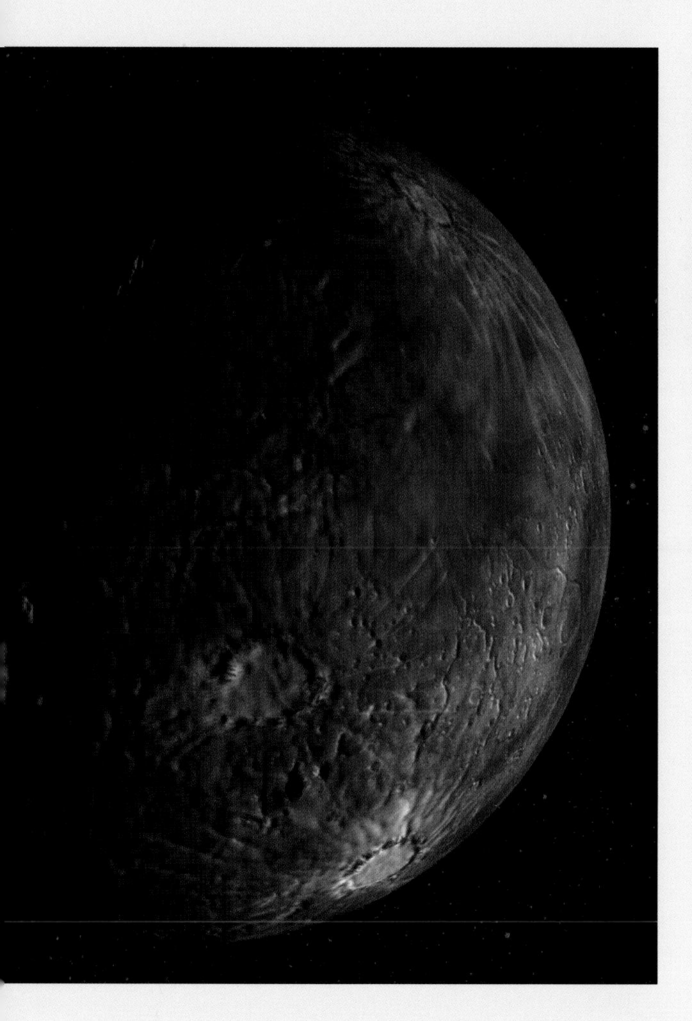

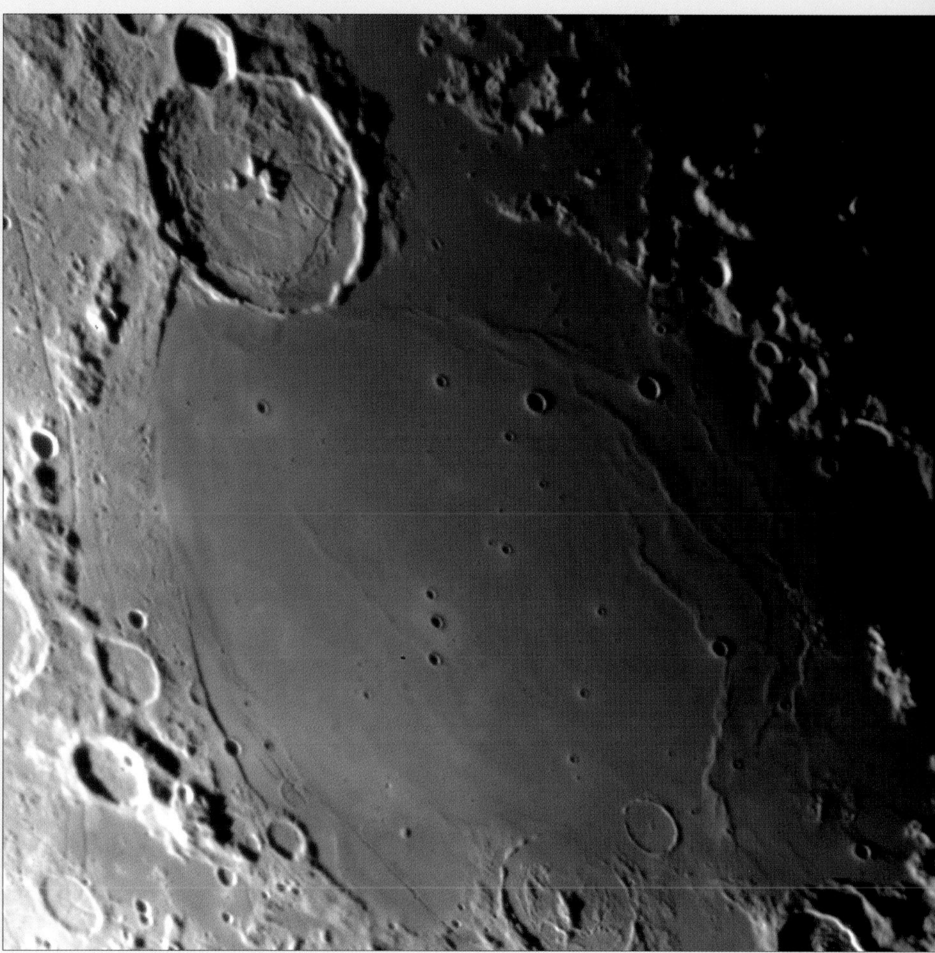

✧ One day, in these richly chequered planetary systems created from the waste products of earlier generations of stars, if all the initial conditions are favourable and all the evolutionary circumstances co-operate, single-cell organisms may be created, biological complexity may increase and life may flourish. The miracle of complex life, consciousness and intelligence will later take place on coagulated rubbish, accumulated grit, cold ash and cinders from the birth of a star. At least, as long as it gets the chance. Because nature does not care; the Cosmos is indifferent. If migrating giant planets turn a young Solar System upside down, if gravitational disturbance scatters planets around like confetti, there is not one force of nature that will protest that life in the system has had no chance. The evolution of the Universe has delivered the building blocks but there is no architect with a blueprint. Either the building constructs itself or it is not constructed at all. Constantly surprising, constantly different. Chaos, chance, imperfection.

✧ By now, the Cosmos is nine billion years old. It has expanded and cooled; evolved from a hot, dense porridge of radiation and particles to a bleak, dark vacuum. Tendrils of mysterious dark matter stretch out like a three-dimensional spider's web. Atoms are gathered into countless galaxies drifting like islands of flickering starlight in this black ocean. On the edge of one of these cosmic archipelagoes, spinning slowly, is our Milky Way – a gigantic flat disk of hundreds of billions of stars and many thousands of star clusters, gas nebulae and molecular clouds. In

the centre of the disk, where the stately spiral arms seem to originate, the galaxy is denser, the stars are closer together and revolve more quickly, driven by the gravity of a monstrous black hole lurking in the core of the system.

✧ Less than thirty thousand light-years from the core, in the quiet suburbs of the Milky Way, totally inconspicuous among the countless stars in its immediate vicinity, the Sun is born from an accidental condensation of gas and dust. The protoplanetary disk coagulates, condenses and is blown clean; the nuclear fire in the centre of the young star ignites. We now await what is going to happen.

✧ At a distance of billions of kilometres, this pinprick of light is surrounded by a sparse cloud of cold cometary nuclei stretching almost halfway to the next star. These are ice-planetesimals that are thrown out of the protoplanetary disk by the gravity of newly formed giant planets. They drift invisibly around the Sun in languid, sluggish loops. Only very occasionally does one fall towards the Sun to make its appearance in the inner regions of the Solar System as a spectacular comet.

✧ A similar belt of comets lies closer to the Sun: small and large lumps of ice created in the outermost parts of the protoplanetary disk which have never had the opportunity to coagulate into a new planet. And further inwards are four giant planets, each with its own impressive retinue of moons and rings, with clouds and eddies in its atmosphere and small, solid cores of rock, metal and other crystals.

✧ So large and massive is the innermost of the four giant planets – created just outside the snowline of the solar system – that the gravitational disturbances it produces has prevented rocky planetesimals at a slightly smaller distance from the Sun from being able to merge into a larger object. This is the

The impressive tail of a comet (*above*) consists of particles of gas and dust pulled loose from the small, icy comet core (*below*).

Irregular lumps of rock – planetoids – are rubble left over from the initial period of the Solar System.

asteroid belt – a band of rubble left over from the formation of the Solar System and regularly shaken up by Jupiter's gravity.

✧ And if you were to look closely, you would see four pebbles moving around the Sun inward from this belt of grit. Four small, cold planets that shrink totally into insignificance against the fierce blazing of the Sun and the soft glow of the gas giants. These are solidified lumps of magma with cores of nickel and iron, a brittle crust of rock and a thin envelope of poisonous gases. Look! That one there is the biggest: the third rock from the Sun.

✧ Welcome to Earth.

Look! That one there is the biggest: the third rock from the Sun

7

Moulding

Collisions, impacts, volcanic eruptions – even in the Solar System, nothing stays as it was. Venus and Mars are derailed, but the Earth becomes a warm, fertile world of water

ALTHOUGH WE DON'T ALWAYS REALISE IT, the Earth cannot exist without the Cosmos. The ground under our feet, the air we breathe, and the water we drink are all products of a billion years of cosmic evolution. Without this celestial prehistory there would be no Milky Way, no Sun, no planetary system and no Earth. The Earth is one with the Universe.

Neither can the existence of our home planet be considered separately from that of the other planets in the Solar System. The Earth does not exist in isolation and would never have been created without the simultaneous birth of other planets such as Mars, Saturn and Pluto. Just as the Milky Way is one of the countless building blocks of the Cosmos, and just as the Sun is an insignificant part of a galactic ecosystem, so is the Earth an inseparable part of a larger whole. And so it is simply by observing this greater whole, by examining the profuse diversity of planets, that we get a picture of the evolution of the Solar System.

Because the Solar System is constantly moving and changing, each of the celestial bodies undergoes slight or major changes of appearance. Here a moon is pulverised, there a climate runs amok, and somewhere else volcanoes spew forth ash and sulphur. Even the apparently fixed orbits of the planets do not remain steady for eternity; no one can predict exactly where Jupiter or the Earth will be a hundred million years from now.

Perhaps an exception can be made for the trillions of icy bodies in the Oort Cloud – the immense halo of small cometary nuclei that represent the remains of the creation of the Solar System. In the cold depths of the Universe, ice-crystals are inert and chemical reactions rarely take place. Just as rare are collisions or disruptions caused by cosmic impacts: the comets in the cloud are far too distant from each other to allow this. And in an object no more than a few kilometres in size, neither do geological processes get a chance. The Oort Cloud comets are fossilised remnants from the prehistory of the planetary system.

This applies to a lesser degree to the larger objects loitering closer to the Sun in the Kuiper Belt – the diffuse band of ice dwarfs and supercomets extending from the orbit of Neptune to a distance of more than eight billion kilometres from the sun. Despite their icy composition, they reflect no more than five to a maximum of fifteen per cent of the sunlight striking them, as if their surface were veiled by dark, carbonaceous compounds. Chance collisions, unannounced meetings, have formed a great many binary objects here: two ice-dwarfs that waltz around each other in a slow dance of gravity and together describe a wide orbit around the distant Sun.

One of Saturn's moons has been shattered and pulverized into a flat ring that surrounds the planet on all sides.

The largest of these deep-frozen binary objects is the duo Pluto and Charon at an average

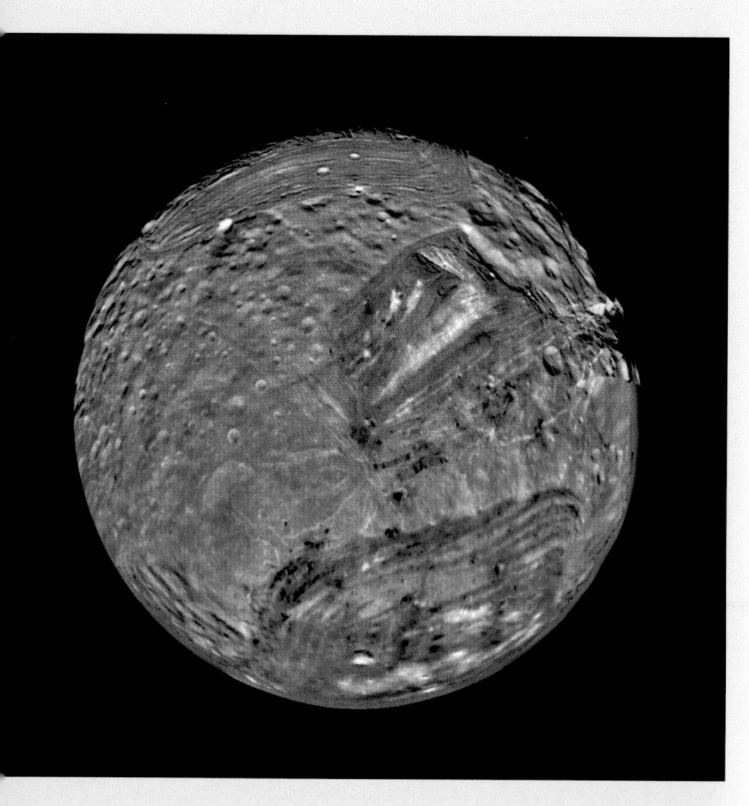

Miranda (*top*) shows the scars of a catastrophic event. The rings of Saturn (*bottom*) were also probably caused by a cosmic collision.

distance of about six billion kilometres from the Sun. Compared with the other ice-dwarfs, Pluto is so large and massive that it manages to retain a thin atmosphere when its elliptical orbit brings it slightly closer to the sun every 248 years and the surface ice partially sublimates. But this dark, cold world on the edge of the Solar System can no more be justifiably called a planet than can the countless other ice-dwarfs in the Kuiper Belt, some of which may be even bigger than Pluto. Pluto is not the Lilliputian of the planetary system but the king of the Kuiper Belt.

Uranus and Neptune – silent, cold gas giants with cores of crystal, mantles of methane and a multitude of satellites – are the outermost of the real planets in the Solar System. Their diameters are roughly five times that of the Earth but they are totally dissimilar as far as composition, temperature and appearance are concerned. Tornadoes and plumes of white cloud blow through the deep-blue atmosphere of Neptune; hazy layers and small storm systems lie hidden in the green-blue atmosphere of Uranus.

Both planets exude an atmosphere of peace and quiet but they also bear the scars of a turbulent past. Neptune has an icy satellite, Triton, with geysers of nitrogen, that revolves around the planet in the wrong direction; with a horizontal axis of rotation, Uranus, with its horizontal axis of rotation, appears to have fallen on its side, leading to very strange seasons. These are the accidental results of a random process of cosmic collisions in the early infancy of the Solar System.

And there are more marks of violence to be found. One of the moons of Uranus, Miranda, looks like a badly fitting, three-dimensional jigsaw puzzle. Old, eroded plains with equally eroded impact craters are suddenly criss-crossed by young, chaotic fault lines and cliffs tens of kilometres high. Miranda was once struck by another celestial object and the Uranian moon broke into pieces. These pieces later remerged into the geological patchwork that now orbits the giant planet.

Catastrophic collisions also produced the ring systems of the giant planets. A small satellite, not too far away from the planet, is struck by a cosmic projectile – another satellite or a drifting comet or asteroid. The countless fragments quickly spread over the full extent of the original orbit. Lumps of rock and ice circle around the planet in a flat disk precisely above the equator

In the wild youth of the Solar System, the Earth and the Moon were bombarded with a rain of cosmic projectiles.

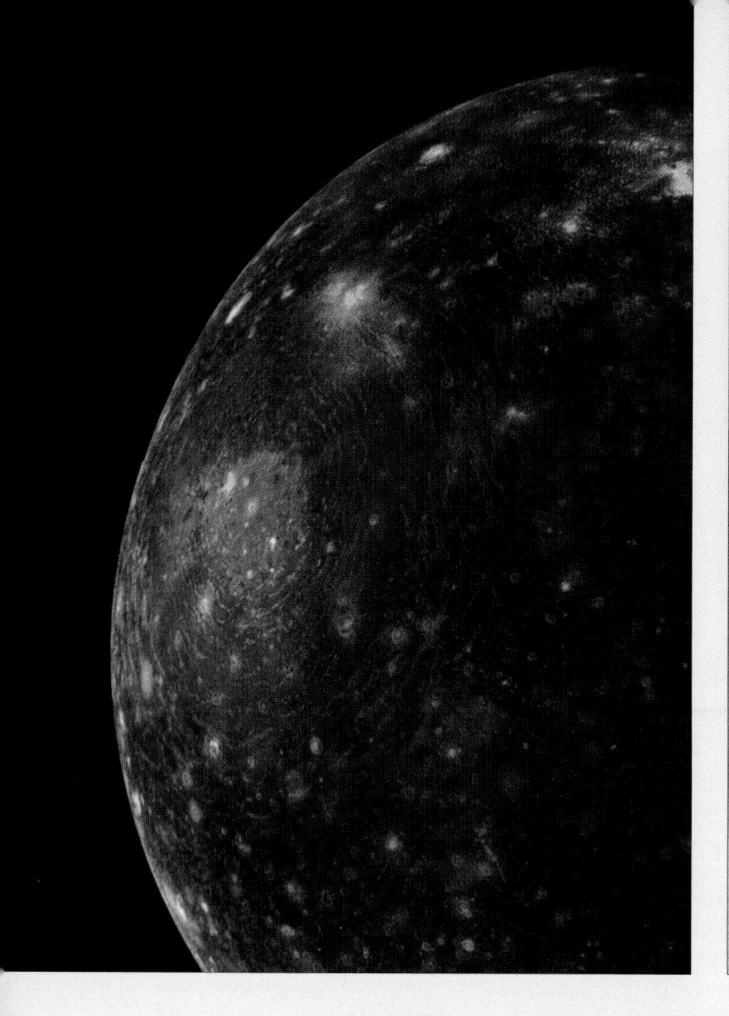

The moons of Jupiter, Callisto, Io, Ganymede and Europa, each have their own characteristic appearance modelled by the forces of nature.

like millions of miniature moons. Subsequent collisions pulverise the material even further. The disk wants to expand inwards and outwards but is prevented from doing so by the gravitational disturbance caused by other small satellites. These gravitational resonances can create the most delicate structures: narrow arcs extending for only a section of the orbit, as in the case of Neptune; thin rings of dark rubble, as displayed by Uranus; an almost structureless disk of dust, as in Jupiter or a glorious, richly detailed ring system like that of Saturn.

✪ At such a short distance from a giant planet, in the direct sphere of influence of strong gravitational forces, the particles in the rings do not succeed in merging into a larger body. The original moon that was broken into pieces by the unfortunate collision never returns. But neither does the ring system live forever. Tiny particles of dust constantly spiral inwards to vanish high in the planet's atmosphere. Planetary rings are temporary jewels with which the giant planets adorn themselves at unpredictable moments.

✪ It is not only the impressive ring system of Saturn that is living proof of the unceasing evolution of the Solar System, for the interior of the planet is also not yet in balance. Like Jupiter, Saturn consists mainly of hydrogen and helium, the two lightest elements in nature from which almost the entire Universe is built. In the planet's centre, pressure and temperature are not high enough to produce nuclear fusion reactions, but the heavy helium nuclei are sinking slowly but surely to the core. This gradual segregation, this slow 'rain' of helium atoms produces warmth

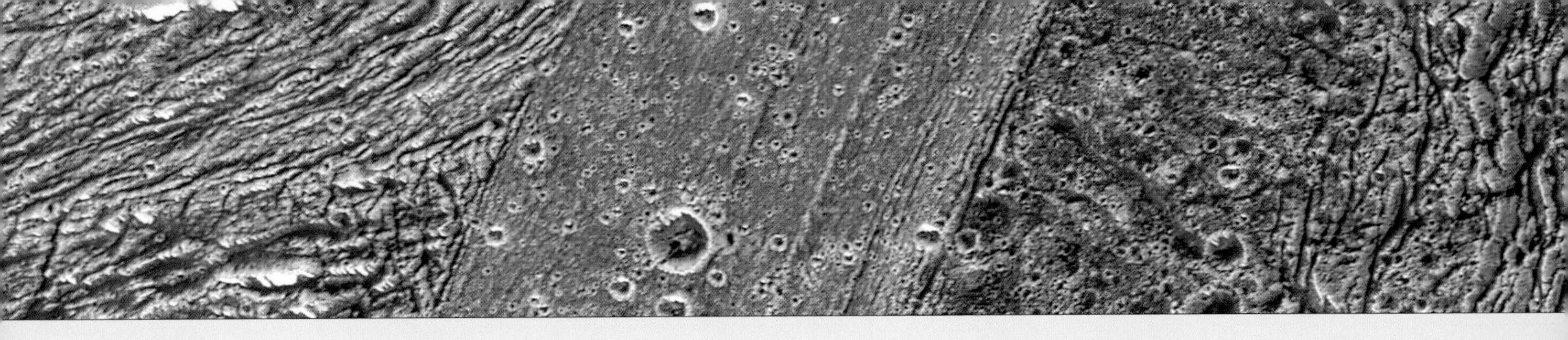

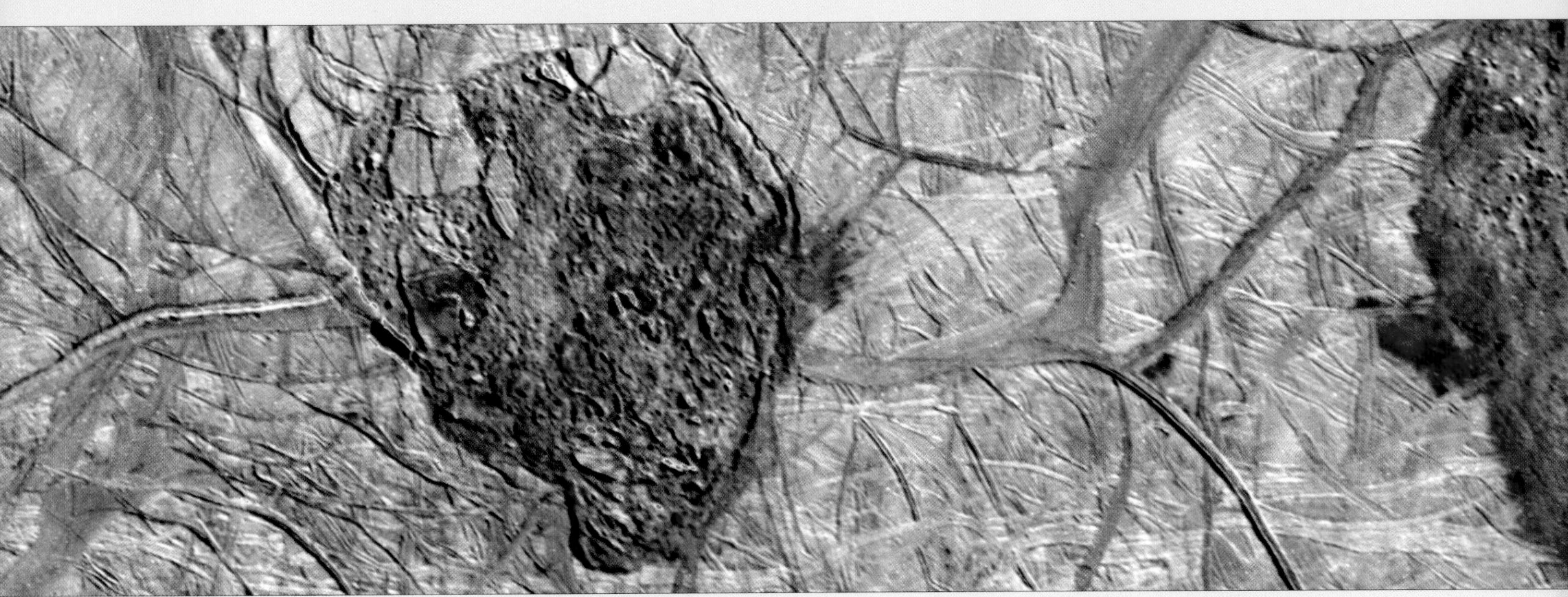

which is radiated forth into space on the outside of the planet. This weak heat radiation is one of the countless indications that the Solar System has not yet fully evolved and that the forces of nature are collectively still seeking their ultimate balance.

Jupiter moves in a much smaller orbit and receives more warmth from the Sun. Because of this, the layers of cloud lie less deep in the atmosphere. Spread out by the fast rotation of Jupiter, they develop into bright and dark belts and zones around the planet, parallel to the equator. At the borders of two cloud belts are small and large eddies that constantly change position, size and appearance, while gigantic flashes of lightning crackle in cumulus clouds stretching kilometres high and auroras make the higher atmosphere glow at the planet's poles.

At great distance from the planet, tens of captured asteroids orbit chaotically. These small moons of Jupiter are barely held in check by the gravity of the giant planet but, closer to this gas giant, four large satellites revolve in neat, circular orbits like pseudo-planets around a pseudo-star. Just as real planets are the by-products of the formation of the Sun, so too have Callisto, Ganymede, Europa and Io been created by accretion in the cloud from which Jupiter was born.

Cosmic impacts, mainly in the early days of the Solar System, have changed the cold surface of Callisto into a desolate rocky desert pockmarked with craters and holes. Jagged mountain ridges cast shadows on a dead landscape that is no longer racked by quakes or volcanoes. Ganymede, a bit bigger and slightly closer to Jupiter, has a more interesting geology with gullies and furrows, old terrain and young fault lines, and countless signs of tectonic activity. Here, the interior is on the move: mountain chains are being thrust upwards and canyons pulled open. Perhaps there is even a form of continental drift.

In a distant past, the Sun may have been reflected in the water of Melas Chasma, a deep valley on the planet Mars.

Europa is the smallest of the four moons but also the most fascinating. Like Callisto and Ganymede, it consists largely of rock and ice, but the strong tidal forces of Jupiter knead its interior to produce sufficient energy to melt the ice. Around the small rocky core lies a mantle of water hundreds of kilometres deep on which drifts a solid crust reminiscent of the pack-ice around the Earth's North Pole, with colliding ice-floes and splitting ice-plates. Sometimes, water from the interior pushes upwards through bursts and splits in the crust; over thousands of years, impact craters lose their shape and are worn flat by the slow, constant working of the ice.

Deep under the surface of this almost mirror-smooth ball are all the ingredients for life: water, energy (supplied by Jupiter's tidal forces) and organic molecules delivered by asteroids and comets that have collided with Europa. Europa's underground ocean, the liquid result of billions of years of evolution, is a potential womb for micro-organisms, a possible cradle for new life.

Contrast this with the hellish world of Io, the innermost of Jupiter's four large satellites! Io is a world of fire and sulphur, of erupting volcanoes and molten rock. Everywhere, its pockmarked surface shows the traces of young and old eruptions: gaping craters, orange-yellow deposits, lava lakes and smoke plumes. The interior, constantly mangled by Jupiter's tidal forces, is one syrupy ball of glowing hot magma. No other body in the Solar System is as volcanically active as Io.

Jupiter has one-tenth of the diameter and a thousandth the mass of the Sun. By cosmic standards, it is an insignificant object – a gaseous ball of fluff, blown around by the 'winds' of

Every rocky body in the Solar System bears the scars of this cosmic bombardment

gravity. But on a more human scale, Jupiter is a colossus that uses its gravity to impose its will on its surroundings. It holds rings and moons in its grip, melts the interiors of Io and Europa, hurls comets through the Solar System and even influences the orbits of other planets. Swarms of small asteroids move in the same orbit as this giant, half of them in front of the planet, the other half behind it, all of them kept in place by the subtle gravitational balance between the Sun and Jupiter. And the giant of the Solar System is also lord and master of the planetoid belt – the wide zone of cosmic rubble between the orbits of Jupiter and Mars. Here it causes resonances resulting in empty zones in the belt, or disruptions that steer the lumps of rock into the inner regions of the Solar System.

This is how the small, Earth-like planets are constantly bombarded with projectiles criss-crossing in oblique, elongated trajectories and crossing or cutting across their circular orbits. Every rocky body in the Solar System – Mercury, Venus, the Earth and Mars, but also the Moon and the asteroids themselves – bears the scars of this cosmic bombardment: craters tens or even hundreds of kilometres in size, produced by the catastrophic impact of these giant meteorites.

The most violent impacts of all took place when the Solar System was very young, in the final phase of the birth process of planets when celestial bodies thousands of kilometres in size collided with each other. The disaster scenarios of this cosmic battlefield can still be

The bare, lifeless Moon, which was once created by a cosmic catastrophe, is the near neighbour of our blue home planet.

reconstructed. For example, Mercury, the innermost planet of the Solar System, was born as a relatively large world with a sturdy core of iron and nickel surrounded by a thick mantle of rock. But a large part of Mercury's mantle was shattered when it collided with a smaller object, and what remained was a smaller planet with a comparatively large metal core.

Venus may owe the unusual orientation of its rotational axis to a similar catastrophe. The planet rotates remarkably slowly and also in the wrong direction, as if it has been turned upside down by a huge impact. On Mars, such a collision may have played a decisive role in the climate change that the planet underwent billions of years ago. The thick atmosphere that the planet once possessed and that may have enabled the origin and early development of microscopic life-forms, largely vanished into space, causing the Red Planet to cool and dry out.

And what about the Earth? Its nearest neighbour in the Universe owes its existence to celestial violence. The silver-white Moon, the symbol of silent peace, is the tangible result of a titanic collision in the riotous youth of the Solar System. The guardian of the Earth is a war-child, born out of a cosmic conflict.

The Earth is still young and hot, recently created from dust sticking together, pebbles and lumps of rock pressed into each other, and colliding planetesimals. Iron and nickel sink to form a slowly solidifying metal core. Molten rock drifts upwards; glowing magma works its way up from the plastic mantle through the brittle crust. The surface of this young planet is a hell of

Radar reveals the surface of Venus, usually hidden below a thick layer of cloud.

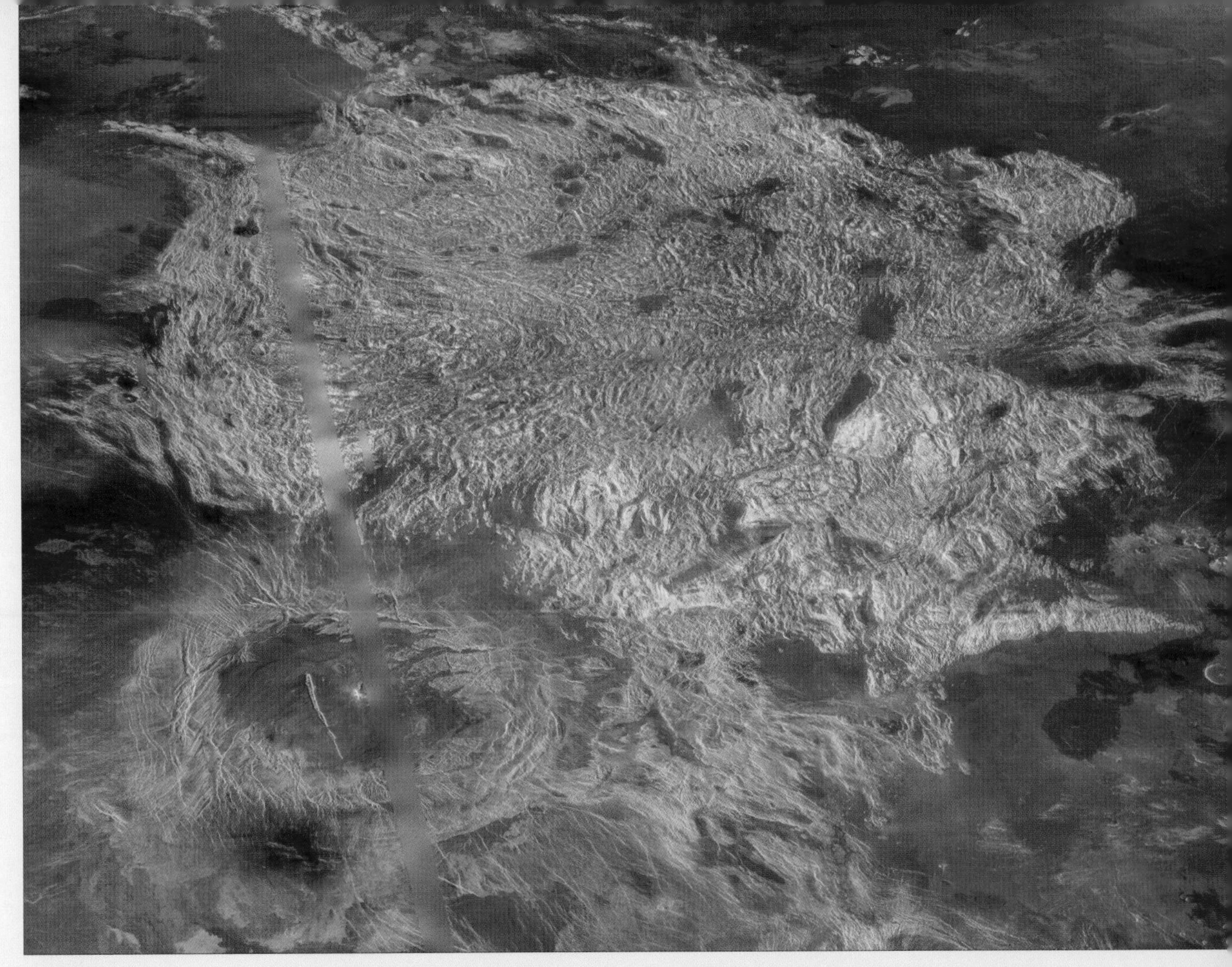

Gullies and flow patterns in the wall of this crater on Mars suggest that there may be water under the surface.

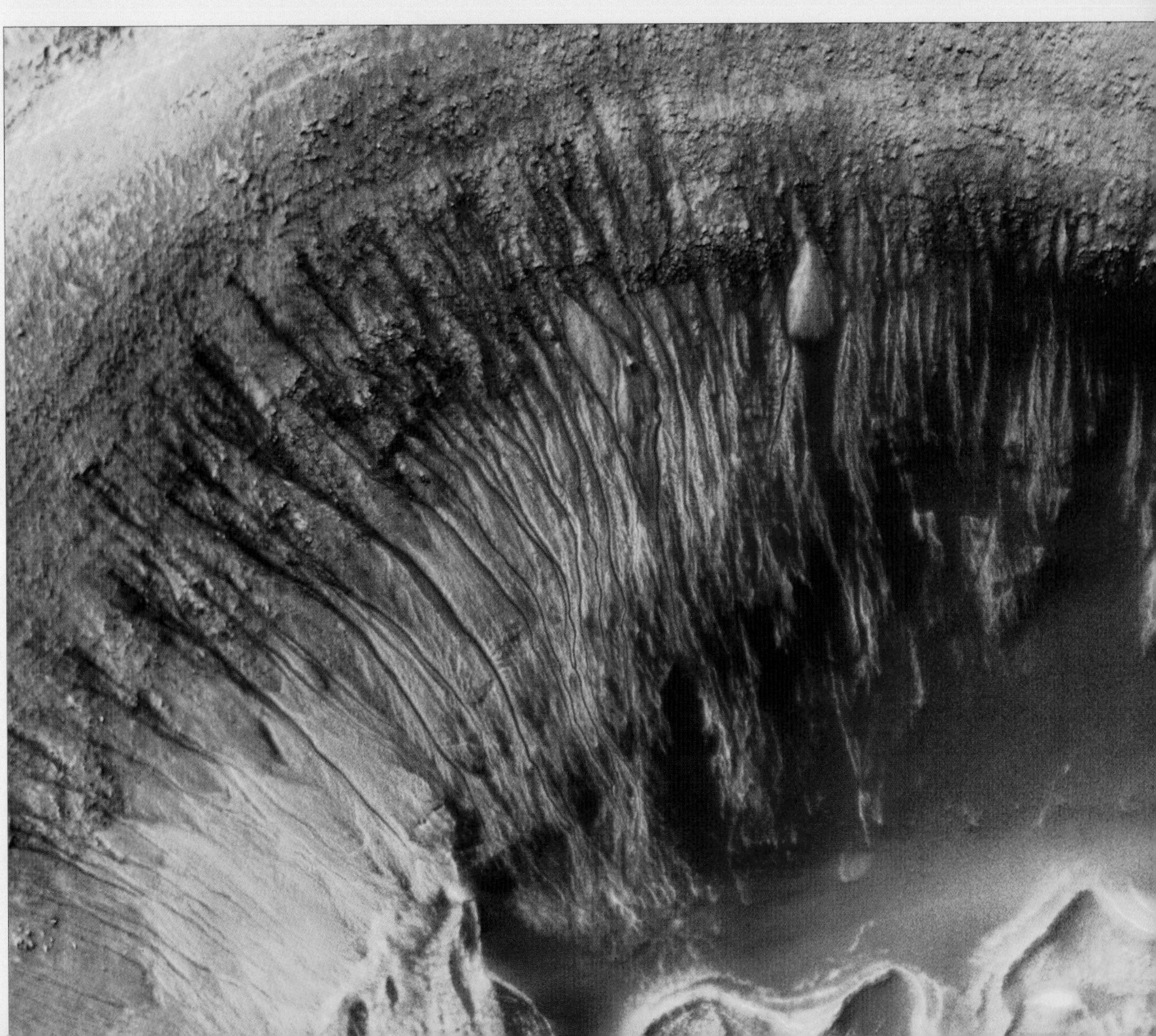

A planet of rock and metal: tiny Mercury is a heavy ball pockmarked by cosmic impacts.

liquid lava flows, boiling volcanoes and simmering pools. A thick mist of water vapour, methane and ammonia, spewed forth from smoking rocks, hangs heavy over the demonic landscape. There is nothing to indicate that this is a world where one day orchids will blossom and hummingbirds will dart from flower to flower.

Unannounced and uninvited, a second protoplanet looms in the distance, just as hot and hellish and with – who knows? – the same potential as the Earth, but lighter and smaller. These two objects have passed each other at a safe distance countless times, each in its own elliptical orbit around the Sun; under the mastery of gravity. But this time a collision is unavoidable. The smaller protoplanet, half the size of the Earth, races to its destruction at a speed of several kilometres per second, obscures the sky above this volcanic landscape, eclipses the Sun and disrupts the boiling ocean and the shimmering atmosphere.

Then the Earth is struck on the flank by this celestial intruder. It shakes and shivers under the violence of the glancing collision. The projectile bursts apart and vaporises as it bores into the planet. Celestial bodies are pulverised and merge; the Earth's mantle spurts liquid rock out into space. So colossal is the energy of the impact that even the Earth's core melts and glows again, while the planet continues to shiver and shake like a quivering jelly.

For many days on end, fire and ash rain from the sky: fragments of the collision fall back to Earth while clouds of dust and cinders obscure the Sun. But a lot of the debris from the collision has been torn loose from the Earth forever. Solidifying lumps of rock swarm in chaotic orbits around a planet still trembling with aftershocks: material from the Earth's mantle, but also remains of the shattered protoplanet. For a while, the Earth seems to adorn itself with a ring-system of cosmic junk but then the inevitable accretion process begins. Gravity imposes its will on matter, rocks merge together and, within a few days following the catastrophe, a new world is created.

So shortly after its birth, the Moon is still shapeless and malleable – a hot ball of crystals and minerals kneaded out of splinters and fragments from two colliding celestial bodies and only a short distance from the Earth. Later, it cools off, its surface becomes ravaged by smaller and larger impacts and, slowly but surely, it recedes from its mother planet and takes on its familiar appearance: an improbably large companion to a relatively small planet with a composition that can be understood only from its turbulent history.

A planet of gas and liquid: the giant Jupiter has no solid surface but constantly changing layers of swirling cloud.

✪ Glancing collisions and cosmic direct hits are run-of-the-mill in the early years of the planetary system, and chaos and chance determine the outcome of the battle. No two planets are alike; every moon tells its own story. Diversity is the name of the game; variation the only constant. And even when the smokescreen of the primordial bombardment has lifted, when the Solar System has settled down and the planets revolve in peaceful isolation around the Sun, even then evolution continues and change lies in wait everywhere.
✪ Venus originally looked like the Earth. There are chains of mountains, volcanoes, rivers and oceans. It is a world of wind and water where geological processes mould the landscape and where it seems only a question of time before the first micro-organisms are created. But Venus is too close to the Sun. And the Sun is gradually increasing in brightness and is warming up the planet even further. The atmosphere holds in the heat like a greenhouse. The surface temperature rises, seas and oceans evaporate and the level of water vapour in the atmosphere rises rapidly. This serves only to increase the greenhouse effect. Venus's climate runs amok and there is no saving it. Ultraviolet light rips apart the water molecules in the atmosphere; hydrogen evaporates into space while oxygen atoms bond with the surface rock. Venus boils dry. Volcanoes continue to spew carbon dioxide into the atmosphere – another greenhouse gas that is now no longer washed out by rain. The atmosphere becomes thicker and thicker; the pressure on the scorched surface rises. And so it is that Venus changes from Earth's twin sister into a nightmare planet with a surface temperature of five hundred degrees centigrade – hot enough to melt lead – and a sulphurous atmosphere ninety times more dense than that of Earth.
✪ Shortly after its creation, Mars too has seas and oceans on its surface. In spite of its greater distance from the Sun, the greenhouse effect keeps the temperature on Mars just above freezing. But Mars is too small. Less heat is produced in the interior of the planet than in the cores of Venus or the Earth. Moreover, Mars radiates this heat more effectively. The

planet cools down and geological activity comes to a stop. The gigantic shield volcanoes, three times the height of Mount Everest, are silenced forever. No longer is the carbon-dioxide supply in the atmosphere supplemented by volcanic eruptions, although the gas continues to be washed out of the atmosphere by rain. The greenhouse effect becomes smaller and smaller and the surface temperature drops. To add to this, when a large part of Mars's atmosphere vanishes into space, possibly as a result of a cosmic impact, the planet's fate is sealed. Mars freezes. The remaining water accumulates in frozen form in the porous planet crust; the atmosphere becomes thinner and thinner. The Red Planet, once perhaps the home of Martian micro-organisms, changes into a cold, dry, rocky desert that no longer resembles the blue oasis of Earth in any way.

Thin, white clouds gather around the highest volcanic summits on Mars. Of all the planets, Mars resembles the Earth closest.

This blue planet has had a lucky escape. It is far enough from the Sun to escape the greenhouse death; large and active enough to resist the cold. While Venus boils dry and Mars freezes, the Earth's surface temperature remains relatively constant, water continues to flow and the planet's geological activity is maintained. It is on this warm world of water, spared by chance from the indifference and capriciousness of nature, that the next step will be taken on the path of cosmic evolution. Here, in the vulnerable twilight zone between boiling and freezing to death, the miracle of complexity comes into being. Here, organic molecules evolve into single-cell microbes. Life is starting on Earth.

At the cold outer edges of the Solar System, blue and silent, floats the giant planet Neptune.

8

Evolution

It has taken the Cosmos ten billion years to create DNA from quarks and electrons on a small planet in a distant corner of the Universe.

The Universe comes to life

A HUMAN BEING IS THE PRODUCT of fourteen billion years of cosmic evolution – a miraculous construction of atoms and molecules, genes and cells, bones and muscles, organs and hormones, body and spirit – but, at the same time, a combination of quarks and electrons that obey the immutable rules of nature. These are particles that were present shortly after the Big Bang, albeit in other combinations; forces that earlier modelled entire galaxies. Now they have come together to form a complex organism that gazes in wonder at the skies and meditates over its own origins.

☆ No one knows if the formation of life is an inevitable event in cosmic evolution. Perhaps it is so. Perhaps increasing complexity will inevitably lead to self-replication and cell division, and micro-organisms are a logical next step following the fusion of elementary particles into atoms and the bonding of atoms into molecules. However, we do not know how the first DNA molecule was created, and when it comes to the formation of the first living cell, science is groping in the dark. So although life may well represent a natural phase in the evolution of the Universe, as a natural phenomenon it is no less enigmatic for all that.

☆ With the formation of the first carbon atoms in the interiors of newly born stars shortly after the Big Bang, the course leading finally to organic compounds and prebiotic evolution was irreversibly set. Once they have been blown out into the cold darkness of space, atoms of carbon, oxygen, nitrogen and hydrogen merge to make organic molecules. In the form of small ice-crystals, these react with interstellar dust particles and, under the influence of ultraviolet starlight and cosmic rays, amino acids are eventually formed – the building blocks of life.

☆ A couple of billion years later, the amount of hydrocarbons in the Universe is already unbelievably big. Complex molecules, elongated carbon chains and organic compounds are now familiar ingredients in the interstellar cocktail. Dark molecular clouds contain sufficient amino acids to people a thousand worlds like the Earth. A little while later, these will be carried down by comets and meteorites to the virgin planets that keep company with the second-generation stars in these clouds.

The living planet: the Earth is the only place in the Universe where life is known to occur.

☆ The building blocks are created during the Big Bang and prebiotic evolution takes place in the vast expanses of space between the stars. But without the security of a planet with warm water and a protective atmosphere, without the nourishing effect of a cosmic womb like the

Earth, all of this would be for nothing. The true miracle takes place in the primeval ocean of a tiny crumb feeding on the light and heat of an insignificant star.

☆ DNA molecules have not been found in interstellar space. Cosmic gas and dust clouds most probably contain no fully-grown micro-organisms. Alien life has not yet been discovered. But in a limitless Universe that teems with prebiotic building blocks and in which a planetary system is a natural by-product of a star's creation process, it is pretty well impossible that life exists on the Earth only. Mars once looked almost identical to the Earth and was hit by the same comets and meteorites. Other stars are also accompanied by planets and whatever happens here almost certainly happens elsewhere too. Even if the origin of life is an extremely improbable process, the Universe is so vast in terms of space and time that it must have taken place more than once.

☆ A single-cell organism is essentially very different to an amino acid or a carbon atom. Prokaryotic bacteria are complete ecosystems of DNA molecules, cell plasma, ribosomes and flagella. These are not terms from the exact world of physics and astronomy but from the much more complex and less well understood world of biology. And yet the carbon atom in the genetic material of a salmonella bacterium does not differ from a carbon atom in the core of a star, and the building blocks of *E. coli* obey exactly the same laws of nature as the quarks and electrons in the Big Bang. Not one molecule in a paramecium displays uncomprehended properties or inexplicable behaviour. It is only at the level of the whole organism that our ability to explain fails us. If it looks like we are talking about an essentially different order, that should perhaps be ascribed to our human perception.

☆ The transition from lifeless molecules and simple replicators to living micro-organisms nevertheless represents an enormous increase in complexity – an act in the drama of cosmic evolution that was possibly unavoidable but yet hardly seemed conceivable ten billion years earlier when the first atoms of hydrogen and helium were created. Compared with this giant leap forward, the further evolution of life from the first single-cellular micro-organisms to the rich diversity of the world around us is a relatively small step, which took only a couple of billion years to accomplish.

☆ Life on Earth is more than three billion years old when the first multicellular organisms appear on the scene. In a sudden explosion of new life-

Dark molecular clouds between the stars contain organic compounds such as hydrocarbons and amino acids.

Around dying stars, matter coagulates to form crystalline materials surrounded with a thin layer of ice in which chemical reactions begin.

forms, the oceans of the Earth become populated by sponges and trilobites, arthropods and vertebrates. Grasses and plants colour the bare landscape green, amphibians leave the shallow coastal waters, and reptiles develop into the lords and masters of terrestrial life. A little later, mammals populate the ecological niches of the Earth's biosphere and not much more than half a billion years after the arrival of the first multicellular organisms – only two per cent of the life of the average star – *Homo habilis* makes hand-axes on the African savannah.

☆ The evolution of life on Earth is no more miraculous but also no *less* spectacular than the evolution of galaxies, stars and planets. In both cases, it's all about a handful of rules and laws that nature is obliged to obey and that lead to an unprecedented and unexpected richness of phenomena: nuclear energy, gravity and electromagnetism in cosmic evolution; mutation, heredity and natural selection in biological evolution. In neither case can the exact outcome of this evolutionary recipe be predicted – no more than the smell and taste of a dish can be deduced from the cookery book.

☆ Darwinian evolution, steered by accidental mutations in genetic material, is unpredictable by definition. Life on Earth does not strive towards complexity, mobility, beauty or intelligence, even though these are qualities that were once missing and now seem indispensable. Evolution has no plan or objective. It gropes around. It tries things out. It goes wrong here and succeeds there. The group behaviour of the ant, the sinuosity of a cheetah, the beauty of the daffodil and the intelligence of a human being are all accidental yet, for all that, are no less miraculous exponents of the Darwinian dice-game. Even if every Earth-like

Micro-organisms were probably created so quickly because prebiotic evolution had largely taken place in interstellar space.

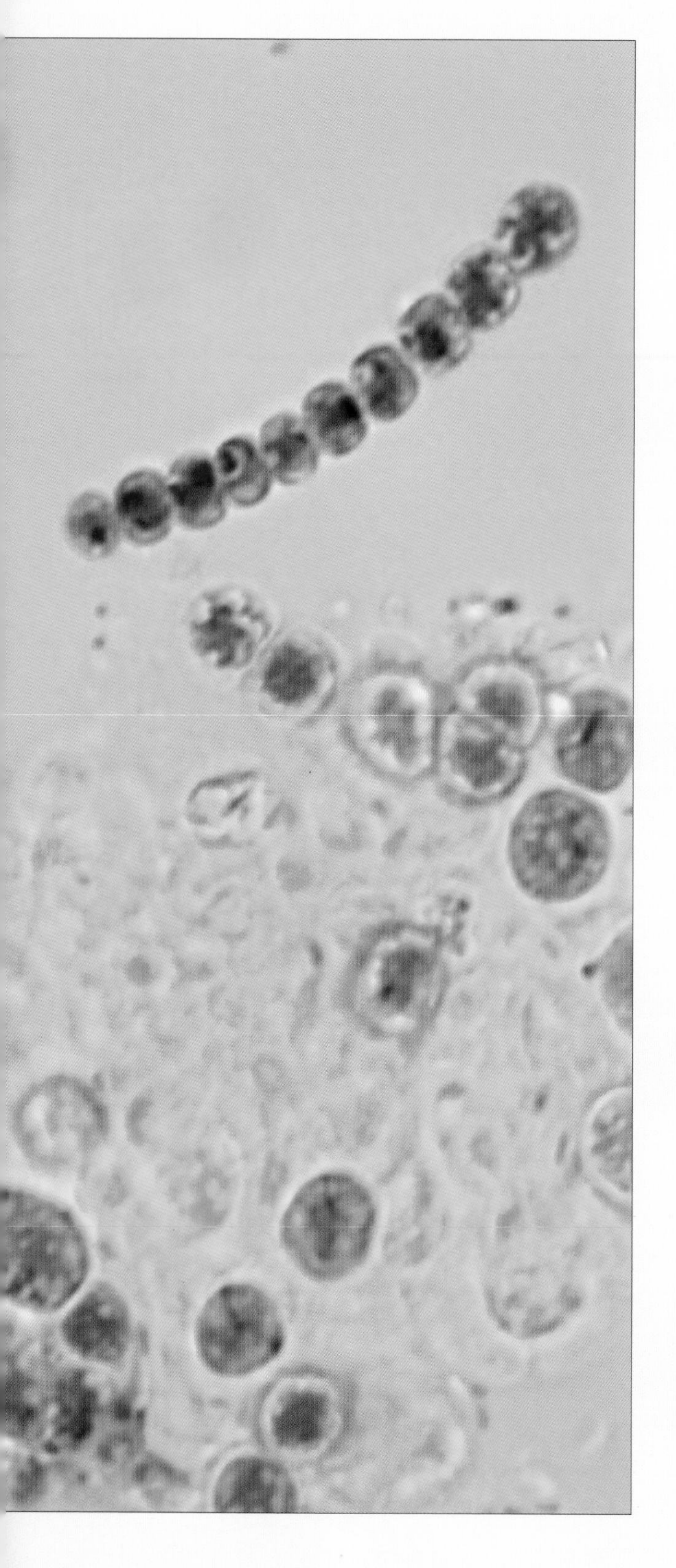

planet in the Universe had started with identical micro-organisms, based on a universal prebiotic building plan, it is still inconceivable that rose petals or peacock feathers would exist on these other planets. Life may well be universal but if you focus on the details, you lay bare the true uniqueness of life on Earth.

☆ Built from elementary particles, ruled by the fundamental forces of nature, created and modelled under the influence of slow, laborious processes, life on Earth is a product of cosmic evolution – an inextricable part of the Universe's diversity. Yet, at the same time, life exerts its influence on this Universe, albeit on a modest scale. The Earth is changing its appearance because of what is taking place in the biosphere. Life is leaving its tracks behind it.

☆ For billions of years, micro-organisms have been producing oxygen molecules that have been accumulating in the Earth's atmosphere. Later, the oxygen level increases even further as a result of the photosynthesis taking place in green plants. From a chemically reducing environment typified by gases such as methane, ammonia and carbon dioxide, the atmosphere changes into an aggressive, oxidising environment that can be maintained only by a constant supply of new oxygen. No single planetary atmosphere in the Solar System is as chemically imbalanced as that of the Earth.

☆ This chemical imbalance in the atmosphere of a planet is probably the best clue in the search for extraterrestrial life. A distant exoplanet, irradiated by the light of its mother star, leaves its individual spectroscopic signature behind in the small amount of radiation it reflects. If you were to succeed in recording and analysing this weak reflection spectrum, you would possibly find traces of an improbable mixture of oxygen, ozone and small quantities of methane gas – a combination that would not exist without biological processes taking place on the planet's surface.

☆ The planet on which we live has been producing another unusual form of radiation for only a few decades now: radio waves on the most diverse of frequencies – powerful carrier waves modulated with clearly artificial signals that partially escape from the ether to spread across the Universe at the speed of light. Searches for similar signals from communicating extraterrestrial civilisations have remained fruitless up to now but, if you do not expect to find rose petals and peacock feathers on a distant planet, perhaps neither should you count on the fact that the Cosmos is populated with life-forms with the same improbable combination of intelligence, consciousness, inquisitiveness, technology and desire for communication as *Homo sapiens*.

☆ More than ten billion years after the Big Bang, when entire generations of massive stars have spewed forth the products of their nuclear fusion into the Cosmos and new solar systems have been created in dark molecular clouds enriched with organic compounds, life on Earth is a fact. Four billion years after this, it is almost impossible that cosmic evolution has not resulted in a capricious, unpredictable biological evolution in a host of places. But anyone who wishes to search successfully for the secluded, remote corners of this immense Universe where life has gained a foothold must not focus his attention on a specific attribute such as interstellar radio messages from intelligent aliens but on the ubiquitous spectroscopic fingerprints of biological activity.

☆ Cosmic evolution is leading from a diffuse, almost unblemished ocean of matter and energy to a hierarchical Universe with a profuse variety of structures on every conceivable scale, from clusters, galaxies, molecular clouds, stars and planets to crystal lattices, molecules, atomic nuclei, nucleons and quarks. Life is located midway on this measuring rod of dimensions – a complex, vulnerable phenomenon composed of the building blocks of the microcosmos and made possible by the cohesion of the macrocosmos. Perhaps the Universe

No single planetary atmosphere in the Solar System

is as chemically imbalanced as that of the Earth

does indeed display an innate tendency towards complexity; perhaps the origin of life is an inevitable consequence of the development set in motion by the Big Bang. But, in spite of this, the Universe shows great indifference where life is concerned because what has been built up in small steps over billions of years can be annihilated mercilessly from one day to the next.

☆ The same processes that are indispensable to the creation of life can wipe life away. Stellar explosions and cometary impacts are the Shivas of the Universe: they bring life *and* death, creation *and* destruction. Even the Sun – the celestial source of light and warmth, the guardian of life on Earth – will be the greatest threat to the biosphere in the future. The Cosmos is no warm, safe environment for something as vulnerable as a fragile living organism. It is an indifferent, inhospitable world: not a Garden of Eden but an Armageddon.

☆ Violent supernovae and gamma-ray bursts turn stars inside out and sprinkle their nuclear ashes over the dark waves of the cosmic ocean. Carbon and oxygen, nickel and iron, silver and gold – the chemical diversity of the Universe exists by virtue of exploding stars and without these celestial fireworks the Earth and life would never have existed. But the stellar explosions also sow death and destruction. Distant planets are sandblasted and scorched by the expanding shells of superheated gases. Powerful waves of X-rays and gamma rays sterilise worlds up to tens of light-years away. Storms of electrically charged particles destroy living cells – delicate biochemical structures that cannot resist this invasion of energetic electrons and protons.

☆ Cosmic impacts are even more disastrous. Frozen cometary nuclei – porous amalgams of

The richness of colours in the sediment of hot-water springs is caused by bacteria that multiply under these extreme conditions.

Sixty-five million years ago, the impact of a comet or planetoid brought the reign of the dinosaurs to an end.

ice and grit surviving from the formation of a planetary system – are the vehicles on which interstellar hydrocarbons, amino acids and other organic compounds arrive on warm, humid, fertile planets like the young Earth. Without this primordial bombardment ravaging the planets in the infancy of the Solar System, the Earth would still be a silent, dead planet as lifeless as the Moon. But comets are also deadly projectiles. They smash craters in the Earth's crust, throw trillions of tonnes of matter into the stratosphere and make the climate run amok. The Sun is hidden for months behind thick clouds of dust, the temperature drops, species become extinct, and food chains are broken.

☆ After the primordial bombardment that took place shortly after the Solar System was formed, the impact frequency fortunately dropped rapidly, yet every hundred million years on average, the Earth is still struck by a cosmic fragment roughly ten kilometres in diameter. This is enough to disturb the climate radically and drastically shift the course of biological evolution. For example, a hundred and forty million years after the appearance of the dinosaurs, their reign was suddenly ended when a comet bored into the Earth's crust at high speed in the Caribbean area creating a crater two hundred kilometres in diameter and wiping out ninety per cent of the species on the planet in a very short time.

☆ Ironically, *Homo sapiens* indirectly owes its existence to this cosmic catastrophe. Without this destructive impact, the giant reptiles would still be in charge on Earth and the fast and varied evolution of mammals would probably never have got started. At the same time, we should realise that the danger of such an impact has in no way diminished and that it is only a question of time before the Earth is once again struck by a natural version of the Last Judgement. In spite of all our scientific knowledge and technical ingenuity, we are essentially subject to the whims of the Cosmos and the caprices of chance.

☆ And yet mankind represents a remarkable and not previously displayed phase in cosmic evolution. There have always been weird celestial objects, complex molecular compounds and surprising natural phenomena: structures and phenomena consisting of the building blocks of the microcosmos and made possible by the cohesion of the macrocosmos. But never before has the Universe been able to observe itself, to reflect on its origins, meaning and future, to ask questions about its own existence.

☆ With the human brain – as far as we know, the most complex object in the Universe – consciousness, self-contemplation and awe make their entrance in the Cosmos. It is as if the Universe gives itself eyes and holds a mirror up to itself. Intelligence and awareness may well be accidental products of an unpredictable Darwinian evolution but they clearly add a remarkable and essentially new element to the already amazing complexity of nature.

☆ Because we are the object of this development, the living witnesses of the Universe's ability to examine itself, we are perhaps not in the neutral position required to assess how curious and improbable this recent phase in cosmic evolution is. However, there is, of course, no one else to ask themselves what the Universe would have looked like if the fundamental constants of nature had slightly different values, if the four basic forces in the Cosmos had more or less differed from each other in strength or range, or if the basic composition of the Universe had been different. At the very least it is remarkable that in all of these cases there would have been no question of stars, planets and complex molecular chemistry, let alone life, intelligence and awareness.

☆ Is the Universe tailor-made for mankind? At the moment of creation, were the forces and constants of nature decided or chosen in such a way that cosmic evolution would automatically result in hierarchy, complexity and intelligence? These are questions for which

Stardust comes to life:
mankind is an integral part
of cosmic evolution.

no one has an answer, riddles for which it is not even sure that a solution will ever be found, and mysteries that apparently lie more in the province of metaphysics and religion than in that of exact science.

☆ Actually, the fact that the unanswered questions of astronomy often appear to have a supernatural component is nothing new. Since time immemorial, celestial phenomena have been linked to the ethereal world of gods and spirits. Cavemen ascribed divine powers to the

Stonehenge was built in the Neolithic era to chart the movement of heavenly bodies.

Sun because it is the guardian of life on Earth. The Greeks and Romans identified the planets in the night sky – Mercury, Venus, Mars, Jupiter and Saturn – with the gods that directed sublunary events and believed that they could predict the future by studying their paths.

☆ However, the accurate measurements and advanced mathematical techniques of Arabian astronomers led to new insights that questioned the assumed central position of mankind and the Earth in time and space. The Polish monk Nicolas Copernicus unleashed a religious riot when he placed the Earth – the dwelling of sinful humankind – on an equal footing with the other planets, that were thought to have a divine perfection. Galileo Galilei, Johannes Kepler and Isaac Newton completed the conceptual revolution and described the fundamental laws of nature that both the fall of an apple and the movement of the Earth around the Sun obey.

☆ The discovery of the telescope marked the beginning of a new revolution in instrumentation. Christiaan Huygens saw polar caps and surface features on Mars; William Herschel discovered a new planet; Friedrich Bessel was the first to calculate the distance to a star. Astronomers charted double stars and nebulae, and mapped the structure of the Milky Way. They discovered that the Universe is expanding. Space travel made it possible to lift instruments above the Earth's atmosphere and actually to visit planets and moons in the Solar System. Satellites and dish antennae enable us to explore the entire electromagnetic spectrum from the longest radio waves to high-energy X-rays and gamma rays. Sensitive electronic detectors and fast digital computers provide scientists with new, unprecedented possibilities.

 Artificial light marks the areas where *Homo sapiens* has changed the appearance of the Earth forever.

By studying the Universe, mankind can also trace his own origins.

☆ It took the Cosmos ten billion years to merge quarks and electrons in a seething sea of radiation into a DNA helix on a small planet in a distant corner of a cold, dark Universe. It took almost another four billion years before life on Earth developed from a single-cell organism to a two-footed hunter who perhaps laid aside his hand-axe to marvel at the beautiful colours of a sunset. One million years later, mankind identified the light of distant stars and, a hundred years after that, explained the history of the Universe. And so we arrive at the present day, at the beginning of the twenty-first century on Earth, at the centre of our own cosmic observational horizon and at the interface of past and future. Here and now, someone is reading a book and stardust is dreaming about his own origins.

☆ But the evolution of the Universe is not over yet. The history of the Cosmos has only just begun. It still has its future ahead of it. And although no one knows what weather it will be next month or when the next cometary impact will take place, the general evolution of the Earth, Sun and Cosmos is known. If we forget about the details and no longer focus on the brief events in one small corner of the Universe but extend our gaze to astronomical proportions, we can leave the here and now behind us and look into the crystal ball of things to come. The future of the Universe starts now.

Leaving Earth. No one knows what the future has in store for mankind.

9

Development

Like a withering flower of light, the Sun finally loses its radiance. But while the Earth petrifies, new stars sparkle elsewhere in the Universe that still have billions of years ahead of them

No more than a couple of centuries ago, people did not look any further than six thousand years back in time. The Cosmos was small and surveyable, the Earth was firmly the centre of the Universe and anyone who wanted to know anything about creation and history consulted the Bible. But thanks to the work of geologists like Charles Lyell, biologists like Charles Darwin and cosmologists like Edwin Hubble, we now know that mankind is part of an evolutionary process lasting billions of years that reaches further back than Adam and Eve and far exceeds the boundaries of our home planet. And anyone who has an eye for the long history of the Cosmos must also direct his gaze towards the distant future.

✦ It is doubtful whether *Homo sapiens* will play a major role in the future of the Universe. We are relatively new arrivals on the stage of life and the way we treat biological diversity, the atmospheric environment and natural resources holds out little hope for a long-lasting presence. Setting aside the way in which we affect our own living environment, we have to observe that in the evolution of life on Earth there have been almost no complex life-forms that have survived for more than a few tens of millions of years.

✦ According to some naïve optimists, science and technology will provide the final answers to overpopulation, hunger and ecological depletion. When the Earth has become too small and uninhabitable, the argument goes, we shall migrate to the Moon, move to Mars or travel to the planets of other stars. Space travel will enable us to colonise the Milky Way, and through a technique known as terraforming we shall control the climate of another planet. We *must* leave the Earth one day because in a couple of billion years the Sun will stop shining.

✦ This attitude is naïve because it ignores the fact that the current population growth can be solved by space travel only if we launch four thousand people per hour into space. Moreover, colonising one, ten or a hundred planets will make little difference to the mathematical laws of exponential growth. And it is completely ridiculous to suppose that mankind will witness the end of the Sun's life. No one can predict the course of biological evolution, but one thing is certain: nothing remains as it is and nature could not care less about the self-importance of man.

A pale, weak Sun shines down on a lifeless Earth five billion years from now.

✦ With the best will in the world, what life on Earth will look like in a hundred, ten thousand or one million years' time cannot be predicted. Probably even the science-fiction scenarios of giant insects, hyper-intelligent super-beings or mindless robots are too conservative. On the other hand, we must realise that, for the first time in the history of life on Earth, molecular

Volcanic activity, earthquakes, plate tectonics and wind erosion change the geology of the Earth constantly. No chain of mountains or ocean is forever.

biology and genetic engineering are enabling us to actively manipulate genetic material. It cannot be discounted that it will soon be possible to influence or steer evolution to some extent.

✦ And yet we cannot escape the conclusion that the major part of the future of the Universe will take place without mankind, just as most of the history of the Universe has taken place without mankind. The expansion of the Cosmos, the life and death of stars and the effects of gravity do not depend on who or what is living on a grain of dust orbiting a bright pinprick. The evolution of the Universe did not suddenly take a different direction with the appearance of mankind and the Cosmos will not shed a single tear when *Homo sapiens* vanishes. The Universe can get on very well without us.

✦ So too can the Earth, for that matter. Our home planet continues to evolve, driven by heat from its interior and by slow currents in the half-molten mantle. Along the Mid-Atlantic Ridge, magma is pushing upwards through cracks in the Earth's crust. The ocean floor is undergoing a ceaseless facelift. Africa and America are drifting apart at a speed of a couple of centimetres a year. The Indian subcontinent is continuing to bore even further into the southern flank of Asia; the Earth's crust is crumpling and the summits of the Himalayas are reaching higher and higher into the sky. Antarctica and Australia are following their own course in this slow continental dance, on their way to new encounters with other plates and fault blocks.

✦ The Ring of Fire – the belt of active volcanoes that surrounds the Pacific Ocean – is spewing and spluttering. The hot spot under Hawaii is producing a new cone, a new shield volcano

possibly even bigger and higher than Mauna Kea. In the Aleutian archipelago or the Philippines, an entire island disappears in the super explosion of a volcano that has remained dormant for centuries. To the south of Iceland, new volcanic islands are rising out of the ocean.

✦ Sideward movements in the Earth's crust cause faults and earthquakes. The San Andreas Fault stretches and releases its pressure; California trembles and shakes. The Great Rift Valley splits East Africa further apart. Turkey and Greece rumble, Japan and China shake from side to side, the Andes heave upwards. Giant quakes are rarer but less innocuous. Perhaps an island slides into the sea or Los Angeles vanishes into the ocean. Tsunamis kilometres high wash over the Earth and batter the coasts of the world's seas. No one is safe from the violence of geology.

✦ And the violence of nature does not just come from the inside. Grains of dust and pebbles burn up in the Earth's atmosphere. Lumps of rock metres in size tumble downwards, melting and hissing, to end up on the Earth's surface as burnt-black meteorites. Larger projectiles create so much air-resistance that they explode at altitudes of tens of kilometres. Less porous pieces of space-rubble blast flaming paths downwards and blow craters in the Earth's crust. A forest is razed to the ground; a city is swept away. An impact in the sea produces immense tidal waves. And the day will come when the Earth is struck by a wandering comet or a drifting asteroid – the day on which a continent will be destroyed and the evolution of life on Earth will certainly make a drastic change of course.

✦ Before that happens, our home planet will go through entirely different metamorphoses. Because of the increasing greenhouse effect, glaciers melt, the pack-ice around the North Pole gets thinner, and the cliffs and icebergs of Antarctica will slowly but surely vanish into the ocean. The Earth's temperature increases, the sea level rises, the Gulf Stream changes course. And yet this global warming will probably be only temporary because the slow, periodical changes in the Earth's orbit and slow fluctuations in the orientation of the rotational axis will soon cause a new ice-age, in the steady rhythm that they have followed already for hundreds of thousands of years.

✦ Warm and cold periods alternate with each other; the sea level rises and falls, and with the constantly changing position of seas, continents and mountain chains, the circulation pattern of atmosphere and ocean changes also. From time to time, the terrestrial thermostat runs amok and the climate is rarely stable for more than a couple of thousand years in succession. Yet there is a noticeable long-term trend: the average temperature on Earth is very gradually increasing because the Sun is slowly getting hotter and brighter. Finally, the fate of the Earth lies in the hands of its mother star.

✦ For billions of years, the interior of the Sun has been producing energy through the fusion of hydrogen into helium. In these nuclear reactions, four hydrogen atoms are converted into one helium atom. Although helium atoms are heavier than hydrogen atoms, the number of nuclei in the Sun's core is gradually diminishing so that the core is slowly but surely getting smaller. In turn, this leads to higher pressure and temperature, causing the outer layers of the Sun to swell slightly and its surface brightness to increase.

✦ This does not happen quickly. It takes one billion years for the brightness of the Sun to increase by ten per cent. But that is more than enough for an irreversible climatic catastrophe on Earth. The average temperature on the planet surface rises to fifty degrees, the polar caps melt for good and the oceans begin to evaporate. The amount of water vapour in the atmosphere gradually increases, taking the greenhouse effect to an unprecedented level and causing the surface temperature to rise even further. While Mercury gets roasted and the hell of Venus is stoked even higher, the Earth changes into a planetary sauna.

✦ This greenhouse drama unfurls so slowly that life on Earth has plenty of time to adapt; the mechanism of Darwinian evolution continues to do its work unhindered. But in the course of billions of years, life gets increasingly less room to maneouvre. The self-reinforcing greenhouse effect eventually leads to the evaporation of all the water on the planet. Under the influence of ultraviolet light, water molecules in the atmosphere disintegrate into separate atoms of hydrogen and oxygen. The

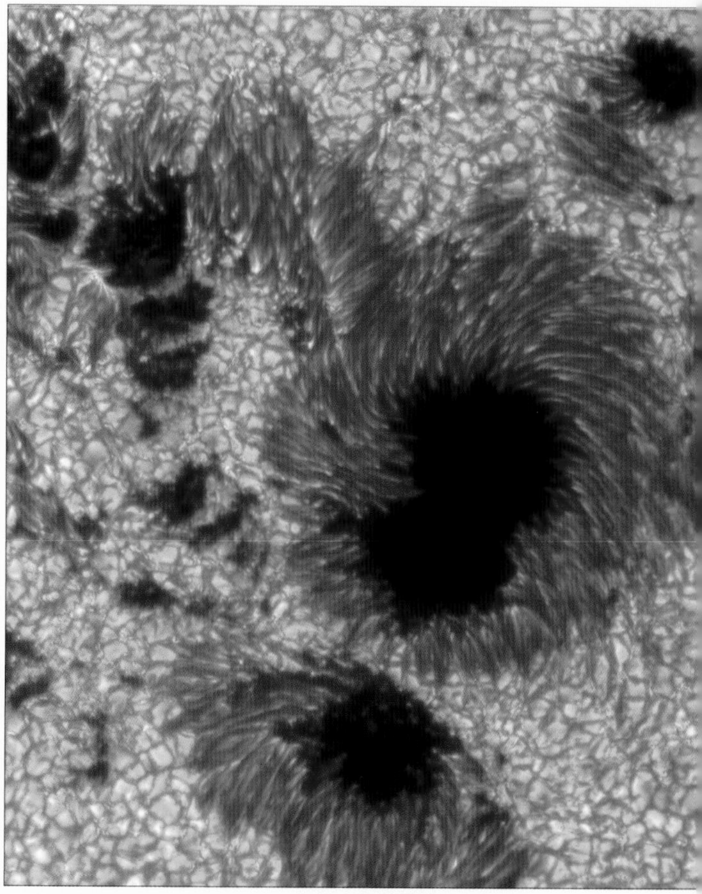

Dark spots on the surface of the Sun are caused by powerful magnetic fields.

Eruptions on the Sun (*left*) are not a threat to life on Earth, but when the Sun changes into a red giant (*right*) our planet will melt and burn to a cinder.

light hydrogen atoms fly off into space; the oxygen atoms are absorbed by surface rock. All of the water that made our world so unique and special for billions of years vanishes almost literally like snow in the Sun. As the temperature on the surface continues to rise unabated, the Earth boils dry.

✦ In the interior of the Sun, the pressure and temperature finally rise to the point at which the helium atoms also start to fuse into even heavier carbon nuclei. In turn, carbon atoms fuse into oxygen. The Sun now consists of a core of oxygen atoms surrounded by a shell in which carbon burning is taking place. Outward of this is a thicker helium-burning shell surrounded by an even thicker outer shell of fusing hydrogen atoms. The nuclear solar inferno produces enormous amounts of energy that blast a path outwards through the hot interior. The gaseous mantle of the Sun grows ever hotter from the inside out, swelling and pushing the glowing surface outwards against the pull of gravity.

✦ The Sun grows horrendously fat. The outer layers become thinner and thinner. And although the Sun is producing more energy than ever before, this energy is spread over a much greater surface so that the temperature drops to no more than 3,000 degrees. For billions of years, the Sun radiated a blinding yellow-white light but now it turns red, almost as if it were ashamed of its size. A bright but cool red giant now resides in the centre of the Solar System – first fifteen, then fifty, then a hundred-and-fifty times bigger than before.

✦ Like a shimmering explosion cloud in slow motion, the fireball of the Sun devours everything around it. The little planet Mercury, already burned and roasted at an earlier stage, is swallowed up by the licking flames of the Sun's surface. Rock becomes liquid, metal melts and the planet vaporises. A little while later, it is the turn of the much bigger Venus. Earth's sister planet vanishes into the solar oven as a glowing ball of boiling lava. By now, the surface of the Earth itself is a boiling ocean of molten rock. The atmosphere has disappeared, the planet steams and reeks, and there has been no sign of life for a long time.

✦ But while the Earth is on the verge of being consumed by its mother star, a small miracle is happening elsewhere in the Solar System. The polar caps of the small, cold planet Mars melt; the

thick layer of permafrost in the soil becomes liquid. Water vapour in Mars's atmosphere causes a greenhouse effect. The surface temperature rises. The atmosphere changes composition and density; the planet gets a tropical climate. After billions of years deep-frozen, water is finally flowing on Mars again and the Red Planet changes into a warm, humid paradise.

✦ Further outwards, Titan awakes – the giant among the planetary satellites. In its orbit around the ringed planet Saturn, Titan was always a cold world with seas and lakes of liquid methane. But Titan is bigger than Mercury and has sufficient gravity to retain a thick atmosphere. It is an atmosphere consisting mostly of nitrogen with smaller quantities of argon, methane and carbon dioxide – precisely the composition of the Earth's atmosphere in the early days of the Solar System.

✦ Organic compounds including ethane and hydrocyanic acid abound in Titan's atmosphere: thick, dark layers of smog created under the influence of ultraviolet sunlight. There is also water and ice and, when the Sun expands into a red giant, Titan changes into a fertile world with all the ingredients for the origin of micro-organisms. After the sterilisation of the Earth, perhaps life in the Solar System will get a second chance in Saturn's back garden.

✦ The metamorphosis of the Sun changes the Solar System forever. Mercury and Venus disappear into the boiling sea of fire. The Earth and the Moon escape but the intense heat of the bloated giant star transforms them into charred cinders. Mars steams and smokes; the thick atmospheres of Jupiter and Saturn expand with the rise in temperature. Europa's ice-crust becomes thinner. Titan begins to blossom. Even in the cold outer reaches of the Solar System, frozen comets begin to steam. The Sun is feverish; its surroundings are clammy. Here a star is fighting its death-struggle.

✦ The interior of the dying star vainly seeks a new balance between gravity and gas pressure. Fusion reactions in its centre succeed each other too quickly; the thin outer layers expand and then collapse again. Expansion and contraction, heating and cooling succeed each other at a constantly changing speed. The Sun shakes and quakes; the colossal ball of gas pulses like a slowly beating heart and the weak gravity on the surface is unable to prevent large quantities of gas from streaming into space with every heartbeat.

✦ There is no longer any stopping it: the Sun is emptying, becoming lighter. The thin outer layers are blown into space. The

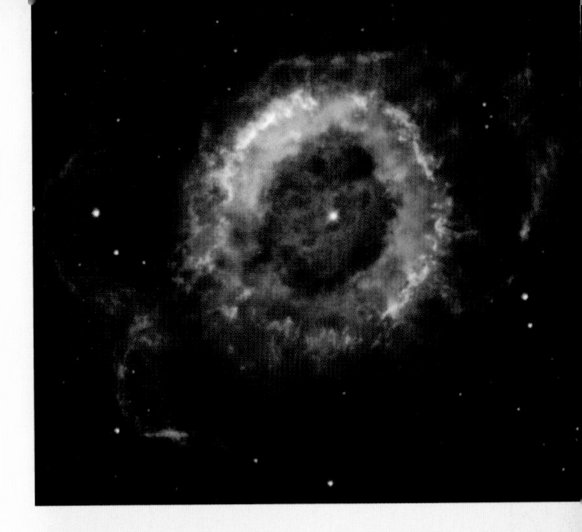

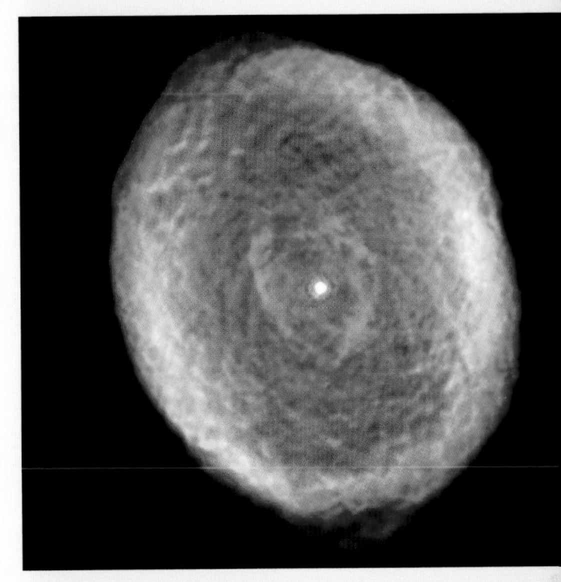

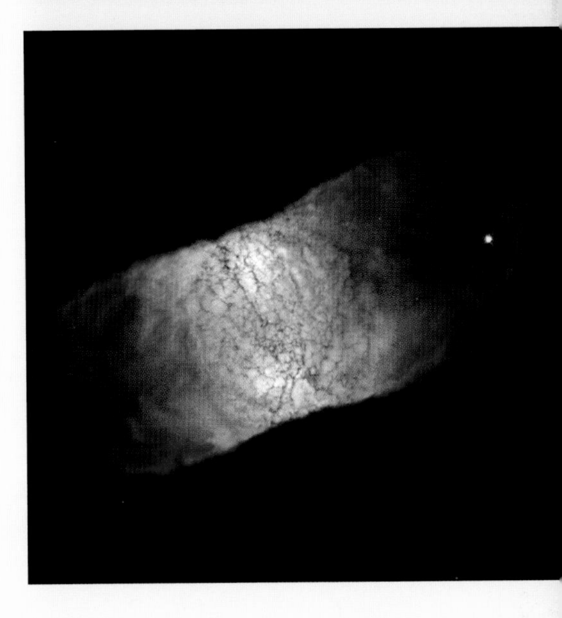

At the end of their lives, stars like the Sun shake off their outermost layers of gas and clothe themselves in a magnificent shroud that slowly blows away and evaporates.

In the distant future, when the Earth is scorched, perhaps life will get a second chance on Saturn's moon Titan.

solar wind, once a light breeze of electrons and atomic nuclei, changes into a shrieking southwester – a true storm of atoms and molecules, powerful enough to erode comets and asteroids and to create spectacular polar lightshows in the atmospheres of planets. Billions of tons of solar gas disappear into the Universe every second; the red giant is like a balloon that becomes porous and is slowly deflating.

✦ This drastic loss of mass, this rigorous slimming operation, also has an effect on the dynamics of the Solar System. The planets are held in their orbits by the Sun's gravity and, at the distance at which they are located, their orbital speed is determined by the mass of the central mother star. But now that the Sun is becoming increasingly lighter, all of the planets start to spiral gradually outwards. The deflation of the Sun leads to an expansion of the planetary system. It is to this outward migration, which started at the moment when the Sun began to swell, that the Earth owes its continued existence; if the Earth's orbit had not got wider and larger, the planet would have met the same fate as Mercury and Venus.

✦ The death agony of the Sun marks the end of the planetary system as we know it. But what constitutes an apocalypse on the insignificant scale of planets, satellites and comets is a fantastic, colourful display for spectators at a distance. The Sun adorns itself with a radiant garland of expelled gas like a water lily that flowers profusely one final time before it dies. For tens of

The red giant is like a balloon that becomes porous and is slowly deflating

thousands of years in succession, this celestial flower is preserved by the Sun's mass loss before finally flying off into interstellar space like pollen in the night.

✦ If you were to look further around you, both in space and time, you would see that the Milky Way is one big flower garden in which, at any one moment, thousands of dying stars are blossoming at the same time. While massive stars explode catastrophically to produce quickly expanding supernova remnants, less massive stars vent their last breath at a calmer rate and wrap themselves in a majestic cloud for tens of thousands of years.

✦ Although, when seen through a small telescope, such 'planetary nebulae' look a bit like dimly lit planetary disks, they have the most varied of shapes. Some are perfectly symmetrical and spherical – precisely what you would expect for a gas shell produced by a spherical star – others are extended, hourglass shaped, asymmetrical or amorphous. Rings, dumb-bells, wheels, rosettes – all occur in the cosmic collection of curiosa.

✦ The nebula is shaped by the star's immediate surroundings. A companion dwarf star, an equatorial dust disk or a planetary system hinders the outward flow of stellar gas in the equatorial region, giving the planetary nebula a laced-up wasp-waist. Powerful magnetic fields do the rest: electrons and ionised atoms are electrically charged and blindly follow the rotating field lines of the stellar dynamo like iron-filings around a magnet.

✦ When the star has emitted approximately half of its mass into space, when it has lost its mantle and clothed itself in a transparent shroud, activity in the core decreases. Fusion reactions stop forever and the star collapses under its own weight into a small, compact ball of degenerate matter – densely compacted atomic nuclei and electrons, ordered into a gigantic crystal lattice the size of the Earth but a hundred thousand times denser.

✦ The compressed stellar gas is hotter than ever before. The surface of the tiny dwarf has a temperature of tens of thousands of degrees and radiates a shimmering blue light. The star is no longer really bright – after all, it has shrunk and has only a small radiating surface – but the photons produced by the hot gas are extremely energetic. Small and dim, but super-hot and glowing white, the star ends its life as a white dwarf.

✦ Only long after the formation of the planetary nebula is the dying star becomes compact and hot enough to produce large amounts of ultraviolet radiation. The energetic photons cause the nebula to glow in the same way in which the atmosphere of a planet produces auroras under the influence of penetrating electrically charged particles or the gas in a fluorescent tube glows because fast-moving electrons are passed through it. And because every sort of atom and every type of molecule produces its own characteristic radiation as a result of this ionisation, planetary nebulae display the most impressive colours.

This richness of colours shows that the expanding nebulae do not consist of purely primordial matter anymore. In addition to the omnipresent hydrogen and the equally unavoidable helium, a planetary nebula also contains enormous quantities of carbon and oxygen, as well as silicon, nitrogen and trace amounts of heavier elements. And while the death throes of the star are still in progress, new molecules are already being born in the expelled nebular matter that will finally be the building materials of a later generation of stars and planets: water, carbon dioxide, methane, silicon oxide, ammonia – some in gas form, others in the form of microscopic ice-crystals – as well as smoke and dust particles, grains of dust and even complex mineral crystals. For this is how it always happens in the evolution of nature: birth and death go hand in hand, and every end means a new beginning.

After a hundred thousand years, the lightshow is over. The nebula has been blown away, extinguished and cooled; the vast cosmic ocean is the final destination of cold stellar gas and dark dust. But the small white dwarf continues to shine. Although nuclear reactions no longer

After a hundred thousand years, the lightshow is over

take place in the centre of the compact star, the enormous interior heat is radiated forth very gradually and it will be many hundreds of millions of years before the star cools off, is extinguished and vanishes from sight forever.

Some white dwarfs even have a spectacular career as a supernova ahead of them. They are part of a binary-star system and, after the red-giant stage and the formation of the planetary nebula, they continue to be accompanied by their less quickly evolving partner, which in some cases finds itself in a much smaller orbit than before. Later, this companion begins to swell too, dumping a large part of its outer layers on the white dwarf. There, under the influence of the strong gravitational field, such an enormous pressure builds up that explosive fusion reactions regularly result. However, because of this mass transfer, the white dwarf becomes increasingly heavy and, when it is approximately one-and-a-half times as massive as the Sun, the electrons in the degraded stellar gas are no longer able to resist the colossal pressure of the star's outer layers. Electrons and atomic nuclei are pressed into each other, the core of the star collapses into a small, super-dense neutron star and the white dwarf bursts like a soap bubble.

However, countless white dwarfs go through their lives alone, just like the insignificant white-hot star that, in a previous life, used to be the radiant central body of the Solar System. The Sun is not destined for an explosive future as a supernova. For a couple of billion years it will pursue its slow course through the Milky Way, getting cooler and dimmer at every orbit. And the planets will travel with it in their wide orbits: cold, dark gas giants faintly lit by the insignificant dwarf star, a small red planet that once again is frozen, and a charred, black cinder, long ago

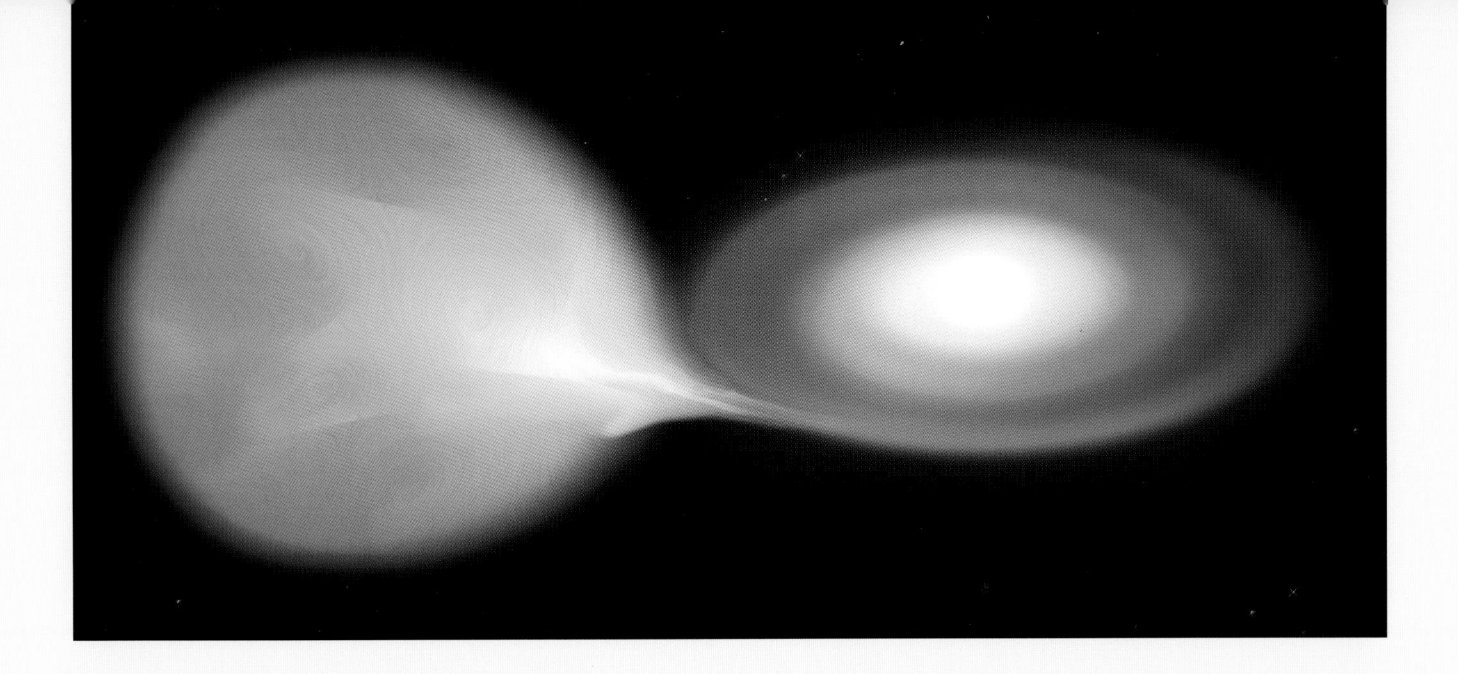

A white dwarf in a binary star system sucks matter away from its companion. If it becomes heavy enough, it will explode as a supernova.

singed and corroded by the inferno of a red giant and now solidified and fossilised into an inert lump of matter without colour or odour – the Earth, once the home of mankind.

✦ Finally the Sun loses its brilliance. The white dwarf cools down so much that it first radiates yellow-white light, next pale orange and eventually dull red. What finally remains is a dark, massive ball of motionless atoms – a black dwarf, the mortal remains of the Sun. A final end has come to the lives of the Sun and the Earth. But just as children's voices sound outside an old-people's home, elsewhere in the Milky Way countless new stars are shining, created from the waste products of earlier generations. Stars that are accompanied by comets rich in carbon and newly born planets, stars that perhaps one day will shine their light on micro-organisms, stars that still have billions of years ahead of them.

✦ It is twenty billion years after the Big Bang. The Universe is still young.

10

Completion

The cosmic clock grinds to a halt in a cold, dark emptiness where space is immeasurable and time freezes. Or will the Multiverse have a new beginning elsewhere?

TO EVERYTHING THERE COMES AN END, and that includes the evolution of the Universe in which we live. We do not know if time will ever come to a stop but what significance does time have when nature itself falls asleep? What is a moment or an eternity when there are no longer events taking place to provide a measuring point in the stream of time? At some time, the Universe will stop evolving, permanency will hold sway and there will be no more change. After that, the Cosmos will solidify, time will freeze and everything will remain as it is, from eternity to eternity.

However, we are a long way from this. Even now, when the Earth is charred, the Sun has lost its splendour and the Milky Way is elderly, the Universe is still young and dynamic. Twenty billion years after the Big Bang, other galaxies are still being formed and new stars and planets are being created everywhere. Supernovae blaze in the darkness, new elements are strewn around the Cosmos and in a host of places the spark of life ignites once again. The development that was set in motion at the birth of the Universe has far from stopped. The Universe still rumbles from the violence of its birth, constantly searching for balance and stability, and the complex wealth of structure on every conceivable scale from picometres to megaparsecs is the tangible result of this quest.

Yet at some point the balance will be found and the Universe will come to rest. The Cosmos expands; galaxies drift away from each other. Isolated systems use up their supply of gas and the creation of new stars comes to a stop. Just as once the first star ignited, so too the last supernova will eventually explode and the last starlight will be radiated. Then there will be a long wait for the complete disintegration of matter, the slow vaporisation of black holes and the everlasting death of the ever-expanding Cosmos.

All of this will take place on unimaginable time-scales against which the present age of the Universe will pale into absolute insignificance, as if we were to attempt to explain the history of mankind to a fly that will live for only a day, or perhaps the creation process of the Grand Canyon to a virtually elementary particle with a lifetime of a femtosecond. And just as human beings can hardly imagine 0.000 000 000 000 000 000 000 000 000 000 000 000 000 000 1 second – the immeasureable age of the Alpha point at which the scientific description of the Universe began – so we can hardly imagine a time-span of 10 000 000 000 000 000 000 000 000 000 000 000 000 years – the age that the Universe will be when atomic nuclei disintegrate and matter vaporises.

In the far, far distant future of the Universe, star-forming will come to a stop and galaxies will evaporate.

Is there really so much time available? Will the future of the Universe last long enough to provide space for these improbable events? Will the present expansion not turn into a contraction much earlier, a universal shrinking that will finally lead to an all-destroying final crunch – a 'Gnab Gib' or reversed Big Bang? Not if you decipher the secret language of the cosmic background radiation, if you break the code of distant supernovae and study the history of the expansion of the Universe. Then it appears that even the gravity of all the matter in the Universe is insufficient to halt expansion, that an uncomprehended vacuum energy, a mysterious repulsive action in the fabric of space-time, is actually making the Universe expand *faster*, meaning that the flow of time has only a beginning but no end and that the future of the Universe is unlimited.

Compared with eternity, twenty billion years is no more than the blinking of an eye. The Universe has sprouted and started to flower but the cosmic springtime is far from over, and autumn and winter are still far away. Countless massive stars have long ago burned up and exploded, even small stars like the Sun have been extinguished and shrivelled, but everywhere in the Milky Way dark clouds of gas and dust are collapsing to form new stars, new suns – third-, fourth- and even fifth-generation stars, rich in heavy elements that find their way into new planetary systems, new chemistry, new life.

A galactic head-on collision taking many tens of millions of years brings with it a tremendous baby boom of new stars. The magnificent Andromeda galaxy, the nearest neighbour

As the Universe continues to expand, cosmic collisions become increasingly rare

of our own Milky Way, has been moving in our direction for billions of years, expanding like a threatening oil-slick over the sky, and now rams the Milky Way in its flank. Stars suddenly have to follow the gravitational marching orders of two galaxies; they are knocked out of their regular orbits, and form dislocated spiral arms and extended tidal tails. The greedy black holes in the cores of the colliding galaxies become entwined in each other's gravitational fields and revolve around each other in a slow dance that finally leads to a scorching merger. But the most immediate effect of the cosmic collision will be a veritable baby boom of new stars caused by powerful shock waves in the interstellar matter. It is as if the galactic vehicles burst into flames after colliding: the birth rate rises by a factor of a thousand and the dark sky above the now lifeless Earth sparkles with the shimmering light of countless new suns.

Of course, the revival does not last long. The most massive stars evolve rapidly, explode as supernovae and fan the fire of star formation for some time to come, but eventually peace returns and Andromeda and the Milky Way merge into a giant elliptical system containing hundreds of billions of stars but almost no interstellar gas. Now it is a question of waiting for the less massive stars with lives of ten or hundreds of billions of years to burn out.

As the Universe continues to expand, cosmic collisions become increasingly rare. Galaxies disappear from each other's sight and sphere of influence. The cycle of birth and death comes to a stop. Although supernova remnants and planetary nebulae still pump trillions of tons of gas into space, some of this always remains locked up in the mortal remains of the stars that emitted it. The yield of celestial recycling is never a hundred per cent. And once the gas supply is exhausted, the formation of new stars comes to an end.

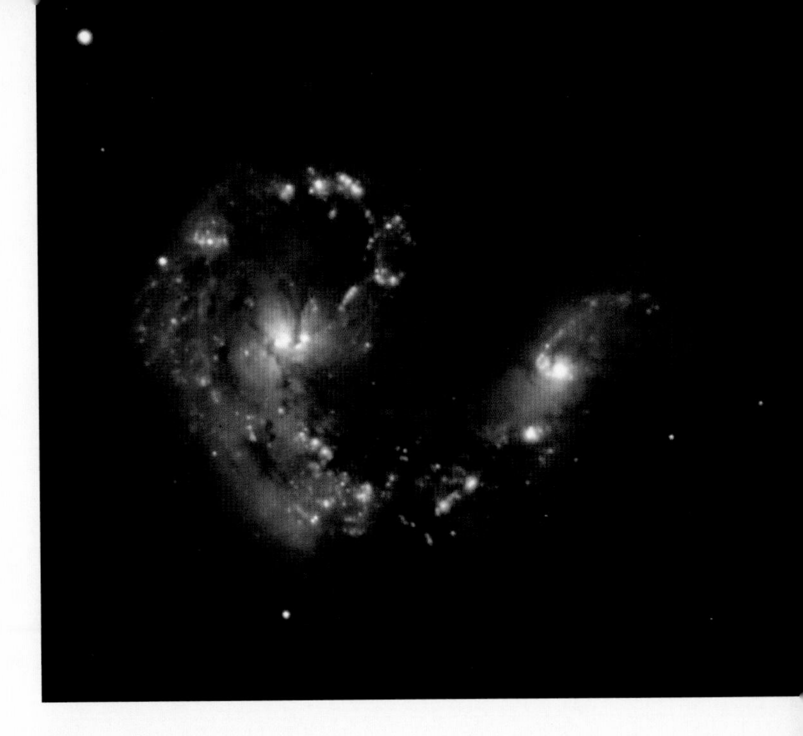

Over tens of billions of years, galactic collisions will continue to relight the fire of star-forming.

When the clock stands at a hundred billion years, the galaxies are populated with insignificant dwarfs and cosmic freaks. The cores of the most massive stars have collapsed into compact neutron stars and black holes. Less massive stars like the Sun have thrown off their exterior mantles and have shrunk into cooling white dwarfs that will eventually lose their colour and splendour and become so cold that they drift through space like black cinders. The only celestial bodies in which the nuclear fire still burns and splutters, albeit on low heat, are the countless stars born small: red dwarfs that are ten times less massive than the Sun and in which nuclear fusion reactions happen so unbelievably slowly that they can continue for many tens of billions of years with even their limited supply of fuel.

Even more numerous are the almost invisible brown dwarfs – objects consisting of hot gas bunched into a ball, scarcely heavier than a giant planet and not hot enough to start the fusion of hydrogen atoms. They wander unnoticed like dark shadows through the vast galaxies that are still faintly lit by the dull glow of the last generation of stars.

But now that time is passing at an increasingly slower rate, now that practically no changes have taken place in the appearance of the Cosmos for billions of years in succession, rare events have a chance – phenomena that are so improbable that they happen no more than once in a hundred billion years. Everything that is not impossible will happen sometime for those who await them long enough. For instance, from time to time, two brown dwarfs will collide with each other and fuse into heavier red dwarfs. Like newly born babies in an old-people's home, these new stars will survive all the other inhabitants of the Milky Way by a wide margin. Very occasionally, even a solitary supernova will flash in the darkness when two *white* dwarfs collide and fuse explosively into a new neutron star.

And yet, the end is simply being postponed. The ultimate disappearance of stars cannot be prevented. When the Universe is a hundred billion years old, the last generation of stars is extinguished forever. Darkness dominates everywhere and the Cosmos is ruled by the gravity that holds dead stars and black holes in its grip. Ironically enough, the same gravity that was once responsible for the formation of clusters, galaxies, suns and planets, now causes the tragic decline of the hierarchical structure of the Cosmos. Over a long period, subtle changes in the motions of the planets bring chaos to the Solar System. The gravity of accidentally passing stars rips planets from their orbits and hurls the fossilised remains of the Earth and the Moon into the dark Cosmos. Mutual gravitational perturbations also occasionally tap stars out of their orderly orbits around the centre of the Milky Way. This exasperatingly slow game of billiards has been going on since the creation of the Milky Way but only over many trillions of years does it have any lasting effect. Planetary systems fall apart, stars quit the systems in which they were once born and galaxies evaporate.

The same gravitational disturbances that hurl stars out into intergalactic space also attract other stars inwards in the direction of supermassive black holes that lie in wait like hungry, underfed monsters, gobbling up all matter drifting into their immediate vicinity. While galaxies evaporate on the periphery, they are consumed from the inside out by black holes that continue to increase in mass. 100 000 000 000 000 000 000 years after the birth of the Universe, even the most impressive galaxies die in this way. The majestic ellipticals and the stately spirals that once constituted the basic building blocks of the cosmic cathedral evaporate and vanish, leaving behind a morbid battlefield of scattered scaffolding, collapsed buildings and dead warriors.

In this dark Universe, dead white dwarfs, super-compact neutron stars and huge black holes wander around. Among these black holes are not only monstrous colossi that once concealed themselves in the cores of galaxies but also small black holes formed by supernova explosions. They are the solitary inhabitants of a gigantic ghost-town. They have not even noticed each other's presence for a very long time, driven apart as they are by the constantly accelerating expansion of empty space.

1 000 000 000 000 000 000 years pass a thousand times; 1 000 000 000 000 000 000 years pass a hundred billion times; 1 000 000 000 000 000 000 years pass 1 000 000 000

Research into distant supernovae shows that the expansion of the Universe will accelerate in the future.

000 000 000 times. Nature's clock no longer ticks in seconds, years or millennia but in an almost endless series of time intervals, each of which lasts a million times longer than the entire radiant history of the Universe. Time seems to lose its meaning and can now be measured only in exponential steps in which each successive phase lasts ten times that of its predecessor.

In this infinite future, when even the memory of stars, planets and life is vague and forgotten, quantum uncertainty begins to take its toll. Protons – the apparently stable nuclear particles that constitute the basic building blocks of all matter in the Universe – appear not to have eternal life after all. Very occasionally, a proton disintegrates into unstable, lighter particles,

The words 'nothing', 'emptiness' and 'vacuum' do not do justice to the current state of the Cosmos

an atomic nucleus loses its identity and a minute quantity of matter is converted into radiation. And even though each of the countless protons in the Cosmos has an average lifespan of at least 100 000 000 000 000 000 000 000 000 000 000 000 years, there eventually comes a point when the loss of mass of the Universe starts to take on alarming proportions.

Protons decay, atoms disintegrate and neutron stars and white dwarfs vaporise. What remains is an immeasurable emptiness in which individual elementary particles – electrons, positrons, neutrinos and dark-matter particles – are further apart from each other than the galaxies were in the era of stars and planets. In this endless sea, large and small black holes bob about like dark gravitational flaws in the fabric of space-time. Their gravitational hunger has not been appeased for a long time now because the world outside their field of vision is empty and black. Yet these sinister objects – once produced by spectacular stellar explosions or from the catastrophic merging of galaxies – are condemned to death and, in time, will also vanish completely from the stage.

Quantum processes in the strong gravitational field just outside the horizon of a black hole produce extremely weak radiation that slowly leeches away into the Universe. Directly contradicting the healthy logic of macroscopic physics, every black hole loses a minute quantity of energy in this way. Normally this loss of energy would be more than compensated for by the slight increase in mass of the black hole, even were it to swallow no more than one atom of hydrogen per century. But now that the nearest particle of matter is billions of light-years away, all black holes are doomed to evaporate sooner or later. They radiate their invisible contents photon by photon. There is simply no other way to keep their weight at the right level.

The evaporation process of a black hole increases in speed as it gets lighter. After an infinite number of eternities, a heavyweight with the mass of a galaxy erodes into a black hole with the mass of the Sun. In trillions of quadrillions of cosmic eons, stellar black holes shrink to become small holes no more massive than planets. These evaporate in a relative blinking of an eye of only some tens of trillions of years into microscopic holes as small as an elementary particle but as heavy as Mount Everest. And then it takes a further ten billion years before such a mini-hole turns inside out and radiates its remaining mass and energy into the Cosmos in an explosion.

Can the forming of a black hole in our own Universe correspond to the birth of a new Universe in another dimension?

When even the most massive of all black holes in the Cosmos has vanished from the scene, the universal clock stands at googol – the official name for a single one with a hundred noughts. Ten to the power of a hundred years have elapsed since the creation – 10 000 000 000 000 000 000 000 000 000 000 000 000 000 000 000 000 000 000 000 000 000 000 000 000 000 000 000 000 000 000 000 000 000 000 years – although the word 'year' has long since ceased to have any meaning. Clusters and galaxies have disintegrated; suns and planets have been dissolved. Matter has given up the ghost and now, with the evaporation of black holes, even the thin fabric of space-time has been ironed smooth. The words 'nothing', 'emptiness' and 'vacuum' do not do justice to the current state of the Cosmos.

Infinitely slowly, nature creeps to the dark point Omega, where space is immense and time freezes.

The evolving Cosmos dies.

Is this the fate of the world we live in? Has the Big Bang brought forth a living Cosmos doomed to die a final death? Is nature actually so morbid that creation and beauty will inevitably have to give way to decline and destruction?

It seems so absurd, so unfair. Of course, life and death go hand in hand at every level in the cosmic hierarchy, but something always follows. Every human life comes to an end but new children are born every day. Mankind becomes extinct but the evolution of life carries on. The Sun and the Earth vanish but organic molecules are combining again in other planetary systems.

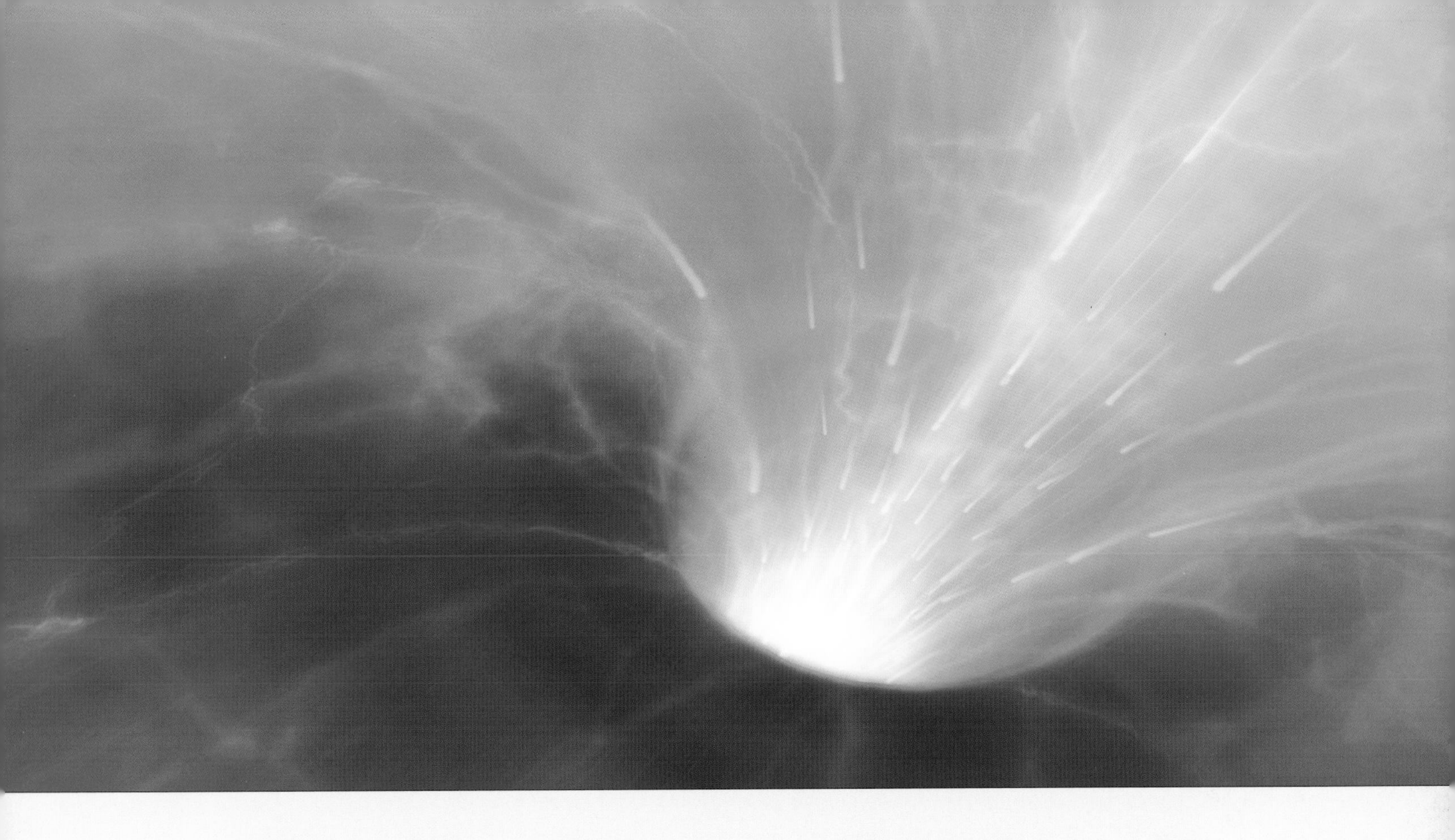

The torch is always picked up and taken further. The end is never final. But there is only one Universe and when the Cosmos comes to a stop, the fire goes out forever.

Or does it?

Let's rewind the film trillions and trillions of years back to the earliest period when the Universe was still young and energetic, when galaxies collided, new stars ignited in dark molecular clouds and the germs of life drifted down onto newly born planets. Stop. That's where we have to be. At the giant star that's on the point of exploding in a spectacular supernova, at the two galaxies entwined in each other's gravitational grip and whose cores are about to merge. This is where the mysterious celestial bodies are created that will resist the stripping of the Universe the longest: the black holes that appear not to obey the laws of nature, that hide themselves beyond the horizon of science.

Is it mere coincidence that every new black hole in the Cosmos – whether it be a relatively small, stellar black hole or a super-massive hole in the core of an active galaxy – seems to form a temporal mirror image of the Big Bang? Just as the First Moment of creation cannot be described by the science of today, for the simple reason that the combination of microscopic dimensions and immensely strong gravity would require a synthesis of quantum physics and the theory of relativity that does not yet exist, neither is the interior of a black hole eager to give up its secrets. And whereas the Big Bang can be typified by the miraculous appearance of matter and energy from an inaccessible singularity, the formation of a black hole is little more than a local collapse of time and space into an all-consuming monster with the same singular properties.

Is there a deeper connection at the basis of this superficial similarity? Is the birth of the

Galaxies, gas nebulae, stars, planets and life – perhaps the history of the Universe is only a short phase in the infinite evolution of the Multiverse.

Universe related to the formation of black holes? Can the appearance of such a singularity in our familiar Cosmos perhaps be synonymous with the creation of an entirely new universe in another dimension?

It seems wishful thinking with a pseudo-scientific basis, but theoretical physicists are taking this possibility seriously. Some mathematical models for the fundamental structure of the Universe even *demand* the existence of a complex network of multiple universes.

From this viewpoint, the future of our own Universe is less tragic than it first appears because, long before the destructive end begins, the Cosmos will already have reproduced itself in the form of countless offspring, each of them just as full of shimmering energy and promising potential as the universe from which they are born. Suddenly, the death of the Universe is no more disastrous than the disintegration of a particular galaxy or the explosion of an individual star – or the death of one ladybird.

And just as our Universe produces an overwhelming number of offspring, there is also an impressive ancestral family tree. Our Cosmos is one of the countless children of the mother universe in which the same cycles of birth and death, creation and destruction take place. This gigantic Multiverse, the genealogical network of cosmic ancestors and offspring, forever separated from each other by impenetrable barriers of unknown dimensions, may well be limitless in terms of size and lifetime – a complex construction of infinite possibilities and eternal evolution.

This is the message of the evolving Cosmos: that nothing is in vain and nothing vanishes unnoticed but that at some time and in some place a new Alpha point will exist – unbelievably small, unbelievably hot, unbelievably compact and unbelievably young. And if you look closely you will see the first fluctuations in density in the expanding primordial matter already and thin clouds of gas collapsing into clusters and galaxies. It will not take long before dying stars scatter new elements, planets will coalesce in swirling disks of gas and dust, and organic molecules will thread together into a living cell.

In this world of mother and baby universes, the ideas of beginning and end lose their

You and I are inseparable parts of this evolving Multiverse

absolute meanings and every event, every movement of an elementary particle, is part of an eternal flux.

You and I are inseparable parts of this evolving Multiverse.

Now close this book and *feel* the flux.

Dust cover ⁓ The Boomerang Nebula, a planetary nebula 5000 light-years away in the constellation Centaurus (The Centaur), photographed with the WFPC2 camera aboard the Hubble Space Telescope. With a temperature of 1° above absolute zero, this nebula is the coldest known place in the Universe. ► ESA/NASA

Half-title page ⁓ ESO 510-13, a galaxy 170 million light-years distant in the constellation Hydra (The Sea Serpent) photographed with the FORS1 camera linked to the European 8.2-metre Antu telescope (VLT UT1). The wavy band of dust is possibly the result of a recent collision with another galaxy. ► ESO

Dust cover and pages 2/3 ⁓ The Eagle Nebula (M16), a star-forming area 4600 light-years away in the constellation Serpens (The Snake) photographed with the 0.9-metre telescope at the Kitt Peak observatory in Arizona ► T.A. RECTOR (NRAO/AUI/NSF AND NOAO/AURA/NSF) AND B.A. WOLPA (NOAO/AURA/NSF)

Pages iv and v ⁓ The centre of the Dumb-bell Nebula (M27) a planetary nebula 1200 light-years away in the constellation Vulpecula (The Fox) photographed with the WFPC2 camera aboard the Hubble Space Telescope. ► NASA/HUBBLE HERITAGE TEAM (AURA/STSCI) / C.R. O'DELL (VANDERBILT UNIVERSITY)

Facing page 1 ⁓ Artificially coloured bubble-chamber photograph of the tracks made by electrically charged particles in the Big European Bubble Chamber of the European Centre of nuclear and particle physics (CERN) in Geneva. ► CERN

Page 2 ⁓ The creation of the Earth from an illustration in the thirteenth-century *Bible Moralisée*. ► CODICES ILLUSTRES

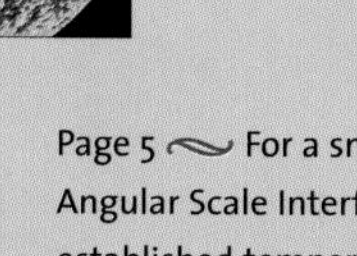

Page 4 ⁓ Projection of the night sky showing minute temperature differences in cosmic background radiation as measured by the American Wilkinson Microwave Anisotropy Probe (WMAP). Blue areas are a minute fraction of a degree cooler than yellow or red areas. ► NASA/WMAP SCIENCE TEAM

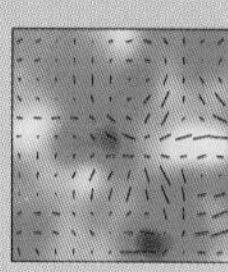

Page 5 ⁓ For a small section of the sky, the American Degree Angular Scale Interferometer (DASI) at the South Pole has not only established temperature differences in cosmic background radiation (shown in colour) but has also measured the direction of polarisation of the background radiation (shown as black lines). ► DASI COLLABORATION

Page 7 ⁓ The evolution of the Universe in a nutshell. The first galaxies were created shortly after the Big Bang. Stars and planets were born in interstellar clouds. Organic molecules carried on comets rained down on planets like the Earth. Life began shortly after. ► NASA/JPL

Page 8 ⁓ Bundles of energy-rich, negatively charged pions trace tracks in a small bubble chamber filled with liquid hydrogen. ► CERN

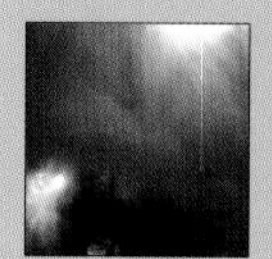

Page 10 ⁓ Young stars and gas and dust clouds near the star R CrA, 500 light-years away in the constellation Coronae Australis (Southern Crown) photographed with the Wide Field Imager on the 2.2-metre MPG/ESO telescope on La Silla, Chile. ► ESO

Pages 12/13 ⁓ Computer simulation of the creation of the large-scale structure of the Universe, with extended superclusters, under the influence of gravity. *Far left*: 13 billion years ago; *far right*: the present-day Universe (the sides of the cube measure 140 million light-years). The simulation is based on a model of the Universe with cold, dark matter and dark energy. ► NCSA/A. KRAVTSOV (UNIVERSITY OF CHICAGO)/A. KLYPIN (NEW MEXICO STATE UNIVERSITY)

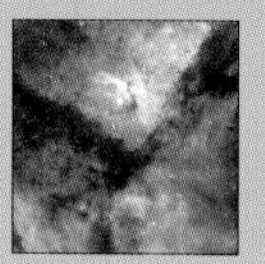

Page 14 ⁓ The Carina Nebula (NGC 3372), a giant star-forming area 8000 light-years away in the constellation Carina (The Keel) photographed with the 0.9-metre Curtis Schmidt Telescope at the Cerro Tololo Interamerican Observatory in Chile. The bright star in the centre is Eta Carinae, one of the heaviest stars in the Milky Way. ► N. SMITH (UNIVERSITY OF MINNESOTA/NOAO/AURA/NSF)

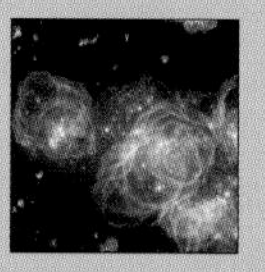

Page 15 ⁓ Artist's impression of the birth of the first stars in the Universe, approximately two hundred million years after the Big Bang. ► A. SCHALLER/STSCI

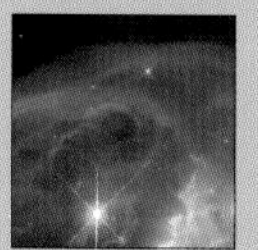

Page 16 ⁓ The Bubble Nebula (NGC 7635) 7100 light-years away in the constellation Cassiopeia, photographed with the WFPC2 camera aboard the Hubble Space Telescope. The star at the bottom of the picture is 40 times heavier than the Sun and is blowing a large bubble 6 light-years in size in the surrounding interstellar matter at a speed of 7 million kilometres per hour. ► NASA/D. WALTER (SOUTH CAROLINA STATE UNIVERSITY)/P. SCOWEN AND B. MOORE (ARIZONA STATE UNIVERSITY)

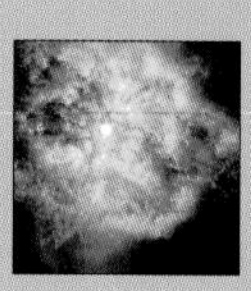

Page 17 ⁓ The giant star WR124, 15 000 light-years away in the constellation Sagitta (The Arrow) has been blowing gas

Abbreviations used:

2MASS = Two Micron All Sky Survey, ACS = Advanced Camera for Surveys, AUI = Associated University, Inc., AURA = Association of Universities for Research in Astronomy, CALTECH = CALifornia Institute of TECHnology, CERN = Centre Européenne pour la Recherche Nucléaire, CXC = Chandra X-ray observatory Center, EIT = Extreme-ultraviolet Imaging Telescope, ESA = European Space Agency, ESO = European Southern Observatory, FOCAS = Faint Object Camera And Spectrograph, FORS = FOcal Reducer/low dispersion Spectrograph, GMOS = Gemini Multi-Object Spectrograph, GSFC = Goddard Space Flight Center, IAC = Instituto de Astrofisica de Canarias, IC = Index Catalogue, IPAC = Infrared Processing and Analysis Center, IRAS = InfraRed Astronomy Satellite, ISAAC = Infrared Spectrometer And Array Camera, JSC = Johnson Space Center, JPL = Jet Propulsion Laboratory, LASCO = Large Angle and Spectrometric COronagraph, M = Messier catalogue, MIT = Massachussetts Institute of Technology, MPG = Max Planck Gesellschaft, MSSS = Malin Space Science Systems, NAOJ = National Astronomical Observatory of Japan, NAOS-CONICA = Nasmyth Adaptive Optics System/Near-Infrared Imager and

into space for ten thousand years at a speed of almost 200 000 kilometres per hour. The resulting nebula, M1-67, contains gas clumps thirty times heavier than the Earth The photo was taken by the WFPC2 camera aboard the Hubble Space Telescope. ► NASA/Y. GROSDIDIER (UNIVERSITÉ DE MONTREAL AND OBSERVATOIRE DE STRASBOURG)/A. MOFFAT (UNIVERSITÉ DE MONTREAL)/G. JONCAS (UNIVERSITÉ LAVAL)/A. ACKER (OBSERVATOIRE DE STRASBOURG)

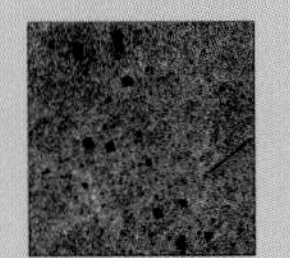

Page 19 (*top*) ⁓ The two-dimensional distribution of approximately two million galaxies over a large section of the southern sky as mapped using the 1.8-metre U.K.-Schmidt Telescope in Australia, based on 185 photographic plates. The density of galaxies is shown in different colours. Black holes have been omitted because bright stars are located there. ► S. MADDOX, W. SUTHERLAND, G. EFSTATHIOU AND J. LOVEDAY (OXFORD UNIVERSITY)

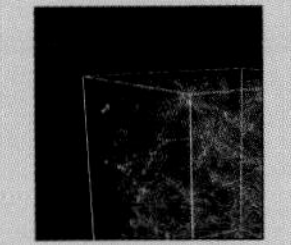

Page 19 (*bottom*) ⁓ Image from a hydrodynamic, three-dimensional computer simulation of the forming of clusters and superclusters, corresponding to the distribution of matter in the present-day Universe. Low-density gas is blue; high-density gas is red. ► M. NORMAN (UNVIERSITY OF CALIFORNIA, SAN DIEGO)/G. BRYAN (NCSA)

Pages 20/21 ⁓ Abell 1689, a heavy cluster of galaxies at a distance of 2.2 billion light-years in the constellation Virgo (The Virgin) photographed by the ACS camera aboard the Hubble Space Telescope. The gravity of the galaxies and the dark matter in the cluster acts as a lens: the light from more distant galaxies is magnified and deformed into bright arcs of light. ► NASA/ESA/N. BENITEZ AND H. FORD (JOHNS HOPKINS UNIVERSITY), T. BROADHURST (THE HEBREW UNIVERSITY)/M. CLAMPIN AND G. HARTIG (STSCI)/G. ILLINGWORTH (UCO/LICK OBSERVATORY)/ACS SCIENCE TEAM

Page 22 ⁓ Thousands of distant galaxies are visible on this photograph of a small section of the sky in the constellation Sculptor, taken with the FORS1 camera linked to the European 8.2-metre Antu telescope (VLT UT1) at Cerro Paranal in Chile. The bright object at the bottom of the photo, to the left under the striking spiral system, is the quasar Q0103-260, at a distance of 11.7 billion light-years. ► ESO

Page 23 ⁓ ACO 3627, a large cluster of galaxies at a distance of 250 million light-years in the constellation Norma (The Square) photographed with the Wide Field Imager on the 2.2-metre MPG/ESO telescope at La Silla, Chile. The yellowish patches of light are the heaviest star systems in the cluster; most of the white dots are stars in the foreground in our own Milky Way. ► ESO

Page 24 ⁓ Ten million stars are visible on this infrared photograph of the centre of our Milky Way at a distance of 25 000 light-years in the constellation Sagittarius

(The Archer). The photo was taken as part of the 2MASS survey using 1.3-metre telescopes on Mount Hopkins in Arizona and at the Cerro Tololo Interamerican Observatory in Chile. At near-infrared wavelengths (between 1 and 2 micrometres) there is practically no absorption of starlight by interstellar dust. ► 2MASS/E. KOPAN/R. HURT

Page 26 ⁓ NGC 4414 is a magnificent spiral galaxy with a bright core and majestic spiral arms of dust and gas at a distance of 60 million light-years in the constellation Coma Berenices (The Hair of Berenice). The photograph was taken using the WFPC2 camera aboard the Hubble Space Telescope. ► HUBBLE HERITAGE TEAM (AURA/STSCI/NASA)

Page 27 (*top*) ⁓ M80 is one of the most condensed globular star clusters in the Milky Way, 28 000 light-years away in the constellation Scorpius (The Scorpion). The photograph was taken by the WFPC2 camera aboard the Hubble Space Telescope. ► HUBBLE HERITAGE TEAM (AURA/STSCI/NASA)

Page 27 (*bottom*) ⁓ A super-heavy black hole with more than two billion times the mass of the Sun in the core of the elliptical galaxy M87, 50 million light-years distant in the constellation Virgo (The Virgin), produces a beam of extremely fast-moving electrically charged particles that cause shock waves in the interstellar gas. The photograph was taken by the WFPC2 camera aboard the Hubble Space Telescope. ► NASA/HUBBLE HERITAGE TEAM (AURA/STSCI)

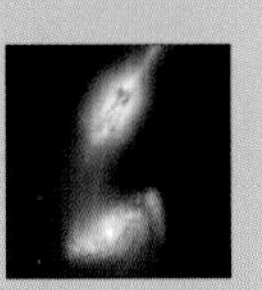

Page 28 ⁓ Two galaxies at a distance of 300 million light-years in the constellation Coma Berenices (The Hair of Berenice) are trapped in each other's gravity. Gravitational forces have drawn long filaments of gas and stars into space, which has earned the duo (NGC 4676) the nickname 'the Mice'. The photograph was taken with the ACS camera aboard the Hubble Space Telescope. ► NASA/H. FORD (JOHNS HOPKINS UNIVERSITY)/G. ILLINGWORTH (UCO/LICK OBSERVATORY)/M. CLAMPIN AND G. HARTIG (STSCI)/ACS SCIENCE TEAM

Page 30 ⁓ X-ray mosaic (approx. 800 × 300 light-years) of the centre of the Milky Way, compiled from tens of photographs from the American Chandra X-ray Observatory. The black hole in the core of the Milky Way is located in the bright white patch in the centre of the photo. Most of the points of light are X-ray binary stars; some of the larger light-sources are the remains of supernovae. ► NASA/D. WANG (UNIVERSITY OF MASSACHUSETTS)

Page 31 ⁓ In the centre of the galaxy NGC 1512, 30 million light-years distant in the constellation Horologium (The Clock) the WFPC2 camera aboard the Hubble Space Telescope reveals a ring of star-forming areas and

Spectrograph, **NASA** = National Aeronautics and Space Administration, **NCSA** = National Center for Supercomputer Applications, **NEAR** = Near-Earth Asteroid Rendezvous, **NGC** = New General Catalogue, **NICMOS** = Near-Infrared Camera and Multi-Object Spectrograph, **NOAO** = National Optical Astronomy Observatory, **NRAO** = National Radio Astronomy Observatory, **NSF** = National Science Foundation, **SAO** = Smithsonian Astrophysical Observatory, **SOHO** = SOlar and Heliospheric Observatory, **SOFI** = Son OF ISAAC, **STSCI** = Space Telescope Science Institute, **TRACE** = Transition Region And Coronal Explorer, **UCO** = University of California Observatories, **USNO** = United Stated Naval Observatory, **VIMOS** = VIsible Multi-Object Spectrograph, **VLT** = Very Large Telescope (UT = Unit Telescope), **WFPC** = Wide Field and Planetary Camera, **WIYN** = University of Wisconsin/Indiana University/Yale University/NOAO consortium.

young star nebulae, with a diameter of 2400 light-years. ► NASA/ESA/D. MAOZ (TEL-AVIV UNIVERSITY AND COLUMBIA UNIVERSITY)

Page 32 (*left*) Artist's impression of two super-heavy back holes in the cores of galaxies, each with its own accretion disk of hot gas, that are about to fuse with each other during a galactic collision. ► CXC/A. HOBART

Page 32 (*right*) Artist's impression of the quasar PKS 0521-36 at a distance of one billion light-years in the constellation Columba (The Dove). In the core of this active galaxy is a super-heavy black hole that sucks up surrounding matter. This produces so much energy that bundles of fast-moving electrically charged particles are blown into space along the axis of rotation of the black hole. ► D. BERRY (STSCI)

Page 33 Explosive phenomena in the core of the galaxy M82, at a distance of 12 million light-years in the constellation Ursa Major (The Great Bear), photographed with the FOCAS camera of the Japanese 8.3-metre Subaru telescope on Mauna Kea, Hawaii. Strings of ionised hydrogen (red) are being blown out of the core at great speed, probably as the result of enormous star-forming and accompanying supernova explosions. ► SUBARU TELESCOPE/NAOJ

Pages 34/35 Computer simulation of the collision of our Milky Way with the Andromeda galaxy, as this may happen in a few billion years' time. The time interval between two successive images is about 90 million years. The two galaxies are first strongly deformed as a result of mutual gravitational attraction. They will finally merge together. ► J. DUBINSKI (UNIVERSITY OF TORONTO)

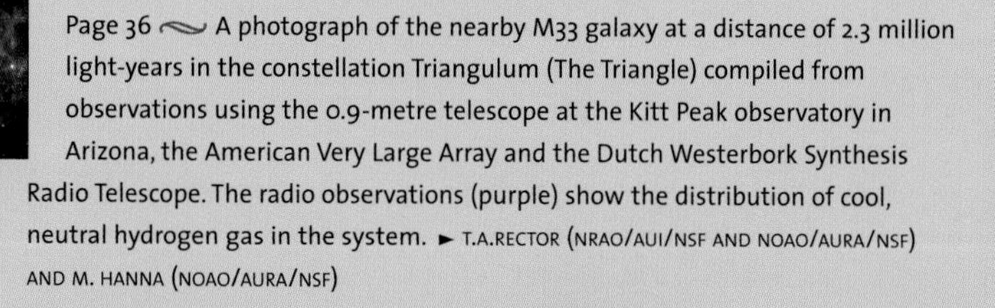

Page 36 A photograph of the nearby M33 galaxy at a distance of 2.3 million light-years in the constellation Triangulum (The Triangle) compiled from observations using the 0.9-metre telescope at the Kitt Peak observatory in Arizona, the American Very Large Array and the Dutch Westerbork Synthesis Radio Telescope. The radio observations (purple) show the distribution of cool, neutral hydrogen gas in the system. ► T.A.RECTOR (NRAO/AUI/NSF AND NOAO/AURA/NSF) AND M. HANNA (NOAO/AURA/NSF)

Pages 36/37 X-ray photograph of the core of the Milky Way taken at the American Chandra X-ray Observatory. The brightest source of X-rays is Sagittarius A*, which coincides with the black hole at the core of the galaxy, at a distance of 25 000 light-years in the constellation Sagittarius. On the Chandra photograph, that was exposed for a total of 164 hours, more than two thousand other weak sources of X-rays have been discovered. ► NASA/CXC/MIT/F.K. BAGANOFF *ET. AL.*

Page 37 These two galaxies in the constellation Hydra (The Water Snake) appear to be colliding but are actually not doing so. The galaxy in the foreground (NGC 3314a) is 117 million light-years distant; the system in the background (NGC 3314b) is 140 million light-years away. The photograph was taken with the WFPC2-camera aboard the Hubble Space Telescope. ► NASA/HUBBLE HERITAGE TEAM (STSCI/AURA)

Page 38 The Rosette Nebula, an impressive star-forming area at a distance of 5500 light-years in the constellation Monoceros (The Unicorn), photographed with the 0.9-metre telescope at the Kitt Peak observatory in Arizona. The colour photo is compiled from three shots. The red shows the presence of ionised hydrogen gas, green shows double-ionised oxygen, and blue, ionised sulphur. ► T.A. RECTOR, B.A. WOLPA AND M. HANNA (NOAO/AURA/NSF)

Page 40 The Orion Nebula (M42) is one of the best known star-forming areas, 1500 light-years away in the constellation Orion. This mosaic of the nebula is compiled from infrared photographs taken as part of the 2MASS survey. The 2MASS survey was carried out using two 1.3-metre telescopes at Mount Hopkins in Arizona and the Cerro Tololo Interamerican Observatory in Chile. ► 2MASS/E. KOPAN (IPAC)

Page 41 (*top left*) Barnard 68, a dark Bok globule at a distance of 410 light-years in the constellation Ophiuchus (The Serpent Holder), is on the point of collapsing into a new star. The countless background stars can still be seen through the cloud on the thin outer edge. The photo was taken with the FORS1 camera linked to the European 8.2-metre Antu telescope (VLT UT1) at Cerro Paranal in Chile. ► ESO/J. ALVES

Page 41 (*bottom left*) The Trifid Nebula (M20), a star-forming area a few thousand light-years away in the constellation Sagittarius photographed with the GMOS camera linked to the 8-metre Frederick Gillett Gemini Telescope at Mauna Kea, Hawaii. ► GEMINI OBSERVATORY/GMOS IMAGE

Page 41 (*right*) The edge of Pelican Nebula (IC5070) approximately 2500 light-years away in the constellation Cygnus (The Swan) photographed with the 0.9-metre telescope at the Kitt Peak observatory in Arizona. The gas and dust nebula (right) has a ragged surface resulting from the ionised radiation of bright, young stars out of shot on the left. ► J. BALLY (UNIVERSITY OF COLORADO)/NOAO/AURA/NSF

Page 43 The Eagle Nebula (M16), at a distance of 4600 light-years in the constellation Serpens (The Snake) photographed with the ISAAC camera linked to the European 8.2-metre Antu telescope (VLT UT1) at Cerro Paranal in Chile. The image is composed of 144 separate photographs in the near-infrared range. At the top right is the open nebula NGC 6611 containing the bright stars illuminating the dark dust columns in the centre. Compare this photograph with that on the title page. ► ESO/M. MCCAUGHREAN

Page 44 HH111 is a protostar at a distance of 1500 light-years in the constellation Orion. The star, shrouded in dust clouds, is at the bottom of the picture (photographed with the NICMOS infrared camera aboard the Hubble Space Telescope); one of the gas jets, 12 light-years in length, can be seen at the top of the picture (photographed with the WFPC2 camera, also aboard the Hubble telescope). ► NASA/B. REIPURTH (UNIVERSITY OF COLORADO)

Page 45 A double nebula (NGC 1850) in the Large Magellanic Cloud at a distance of 170 000 light-years in the constellation Dorado (The Goldfish), photographed with the WFPC2 camera aboard the Hubble Space Telescope. The large star cluster (centre) contains ten of thousands of stars; the smaller one (top right) looks more like the open star clusters in our own Milky Way. The thin strands of gas are probably formed by the explosions of heavy stars. ► NASA/ESA/M. ROMANIELLO (ESO)

Page 46 Stellar explosions in the open star cluster Hodge 301 (bottom right) have caused shock waves in the surrounding gas and dust clouds. Hodge 301 lies in the Tarantula Nebula, a gigantic star-forming area in the Large Magellanic Cloud at a distance of 168 000 light-years in the constellation Dorado (The Goldfish). Photograph taken with WFPC2 camera aboard the Hubble Space Telescope. ► HUBBLE HERITAGE TEAM (AURA/STSCI/NASA)

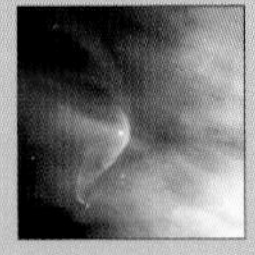

Page 47 The young, changeable star LL Orionis, 1500 light-years away in the constellation Orion, is surrounded with a bow shock caused by radiation and stellar wind from the hot, young stars in the Orion nebula. Photograph taken with WFPC2 camera aboard the Hubble Space Telescope. ► NASA/HUBBLE HERITAGE TEAM (AURA/STSCI)/C.R. O'DELL (VANDERBILT UNIVERSITY)

Pages 48/49 Countless young protostars with their accompanying jets and shock waves (so-called Herbig-Haro objects) are visible on this photograph of the area surrounding NGC 1999, a star-forming area at a distance of 1500 light-years in the constellation Orion. The true nature of the bright 'waterfall' at the top, just right of centre (HH222) is unknown. Photograph taken with the 0.9-metre telescope at the Kitt Peak observatory in Arizona. ► T.A. RECTOR, B.A. WOLPA, G. JACOBY (NOAO/AURA/NSF)/HUBBLE HERITAGE TEAM (AURA/STSCI/NASA)

Page 50 (*left*) The star Merope in the open star cluster The Pleaides (M45), 385 light-years away in the constellation Taurus (The Bull) is approximately 100 million years old and is still surrounded with the remains of the gas and dust cloud from which it was created. Photograph taken with the 0.6-metre Burrell Schmidt telescope at the Warner & Swasey observatory at the Case Western Reserve University. ► NOAO/AURA/NSF

Page 50 (*right*) The Sun is the most studied star in the Universe. It is an average dwarf star with a life of about 5 billion years that was also once created in an active star-forming area. This photograph, on which a gigantic eruption is visible (bottom left) was taken with the EIT telescope aboard the American/European SOHO satellite. ► SOHO/EIT (ESA/NASA)

Page 52 Supernova 1987A exploded on 23 February 1987 in the Large Magellanic Cloud at a distance of approximately 170 000 light-years in the constellation Dorado (The Goldfish). The supernova was visible with the naked eye in the southern hemisphere. Around the exploded star there are now fluorescent rings of expelled gas visible (shown red in this image taken with the WFPC2 camera aboard the Hubble Space Telescope). ► HUBBLE HERITAGE TEAM (AURA/STSCI/NASA)

Page 54 The Halter Nebula(M27), a planetary nebula approx. 1200 light-years away in the constellation Vulpecula (The Fox), photographed with the 2.1-metre telescope at the Kitt Peak observatory in Arizona. ► KITT PEAK RESEARCH EXPERIENCES FOR UNDERGRADUATES PROGRAM/NOAO/AURA/NSF

Page 55 Filaments and bubbles of gas in the nebular complex N44C, created as a result of supernova explosions in a young star cluster in the Large Magellanic Cloud, approximately 170 000 light-years away in the constellation Dorado (The Goldfish). Photograph taken with WFPC2 camera aboard the Hubble Space Telescope. ► NASA/HUBBLE HERITAGE TEAM (AURA/STSCI)

Page 57 Close-up of the Keyhole Nebula, part of the Carina Nebula (NGC 3372), 8000 light-years distant in the constellation Carina (The Keel). The irregular dark clouds (top left and bottom centre) are dust complexes consisting of heavy elements produced in earlier generations of stars. Photograph taken with the WFPC2 camera aboard the Hubble Space Telescope. ► NASA/HUBBLE HERITAGE TEAM (AURA/STSCI)

Page 58 Supernova 1994D in the dust-laden spiral galaxy NGC 4526, 40 million light-years distant in the constellation Virgo (The Virgin), photographed with the WFPC2 camera aboard the Hubble Space Telescope. ► HIGH-Z SUPERNOVA SEARCH TEAM/STSCI/NASA

Page 59 (*left*) A computer simulation of a Sun-like star that has been struck by the shock wave of a nearby supernova explosion (out of the picture at the top of the illustration). The density of the gas is shown in different colours. ► R. HOFFMAN/LAWRENCE LIVERMORE NATIONAL LABORATORY

Page 59 (*right*) Artist's impression of a pulsar – a rapidly rotating neutron star with a strong magnetic field – blowing two powerful beams of electrically charged particles and radiation into space. ► M.A. GARLICK, HTTP://WWW.MARKGARLICK.COM

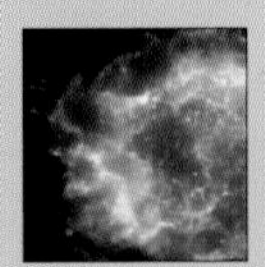

Page 60 Cassiopeia A is the expanded remains of a supernova explosion that took place at the end of the 17th century at a distance of approximately 3400 light-years in the constellation Cassiopeia. This detailed X-ray photograph was taken at NASA's Chandra X-ray Observatory. ► NASA/CXC/SAO

Page 61 The Crab Nebula (M1), the remains of a star that went supernova in 1054, 6500 light-years distant in the constellation Taurus (The Bull), photographed with the 3.5-metre WIYN telescope at the Kitt Peak-observatory in Arizona. The lower of the two brighter stars in the centre of the photograph is the Crab Pulsar, the collapsed remains of the exploded star. ► J. GALLAGHER (UNIVERSITY OF WISCONSIN)/WIYN/NOAO/NSF

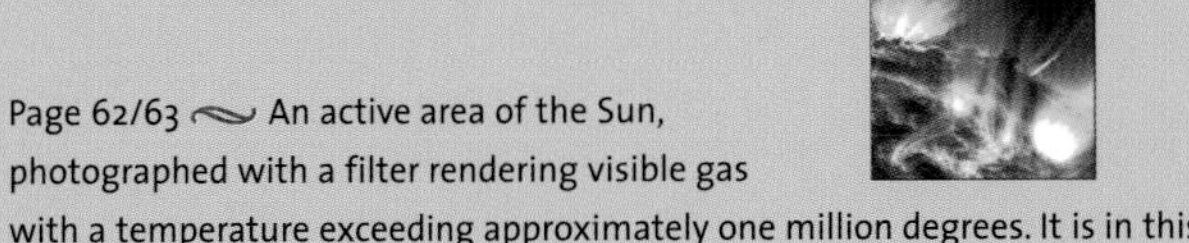

Page 62/63 An active area of the Sun, photographed with a filter rendering visible gas with a temperature exceeding approximately one million degrees. It is in this sort of area that the solar wind originates, a constant stream of electrically charged

particles that are blown into space. Photograph taken with the American TRACE satellite. ► C.J. SCHRIJVER/TRACE TEAM

Page 64 ~ Close-up of glowing filaments of gas in the expanding remains of the supernova Cassiopeia A, the residue left from a stellar explosion that took place at the end of the 17th century in the constellation Cassiopeia. Photograph taken with WFPC2 camera aboard the Hubble Space Telescope. ► NASA/HUBBLE HERITAGE TEAM (AURA/STSCI)

Page 66 ~ The heart of the Orion Nebula (M42), 1500 light-years away in the constellation Orion with, in the centre, the four bright Trapezium Stars ionising the gas in the nebula. This near-infrared mosaic consists of 81 separate photographs taken with the ISAAC camera linked to the European 8.2-metre Antu telescope (VLT UT1) at Cerro Paranal in Chile. ► ESO/M. MCCAUGHREAN

Page 68 ~ Infrared photographs of six young protoplanetary disks at a distance of 450 light-years in the constellation Taurus (The Bull), taken with the NICMOS camera aboard the Hubble Space Telescope. *Top* (*from left to right*): CoKu Tau/1, DG Tau B, Haro 6-5B; bottom (*from left to right*): IRAS 04016+2610, IRAS 04248+2612, IRAS 04302+2247. ► D. PADGETT (IPAC/CALTECH)/W. BRANDNER (IPAC)/K. STAPELFELDT (JPL)/NASA

Page 69 ~ Artist's impression of Gliese 876b, an exoplanet twice as heavy as Jupiter. The planet describes a 61-day orbit around the dwarf star Gliese 876 (Ross 780), at a distance of 15 light-years in the constellation Aquarius. ► NASA/G. BACON (STSCI)

Page 70 ~ Compact globules (Thackeray's Globules) stand out in black against the bright nebular background of IC 2944, 5900 light-years away in the constellation Centaur. The photograph was taken with the WFPC2 camera aboard the Hubble Space Telescope. ► NASA/HUBBLE HERITAGE TEAM (AURA/STSCI)

Page 71 ~ Five images taken from a computer animation of the creation of the Solar System from a contracting cloud of gas and dust. ► ANIMALU PRODUCTIONS/JPL

Page 72 ~ Four phases taken from a computer simulation of the rapid forming of giant planets from instabilities in protoplanetary disks. Differences of density are indicated by different colours. From the simulations, it appears that giant planets can be created in a couple of centuries. ► L. MAYER (UNIVERSITY OF WASHINGTON)

Page 73 ~ Artist's impression of the forming of our Solar System (not to scale). While particles of gas and dust and small planetesimals continue to roam about, some large protoplanets have already been created. ► M.A. GARLICK, HTTP://WWW.MARKGARLICK.COM

Page 74 ~ The giant planet Jupiter with the shadow of the moon Europa on 7 December 2000, photographed by the American space probe Cassini. The image has been compiled from four separate photographs. ► NASA/JPL/UNIVERSITY OF ARIZONA

Page 75 (*left*) ~ Artist's impression of Quaoar, an ice dwarf in the Kuiper Belt at a distance of 6.5 billion kilometres from the Sun. It has a diameter of 1000 kilometres. ► NASA/M. BROWN (CALTECH)

Page 75 (*right*) ~ The 105-kilometre-wide walled plain Gassendi (*top left*), on the edge of the Mare Humorum, a volcanic plain on the Moon, photographed with the Wide Field Imager of the 2.2-metre MPG/ESO telescope at La Silla, Chile. ► ESO

Page 76 (*top*) ~ The long, curved tail of the comet Ikeya–Seki, visible in 1966, slightly above the (tilted) horizon of the Earth, photographed with a 35-mm camera. ► R. LYNDS/NOAO/AURA/NSF

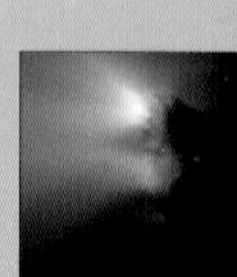

Page 76 (*bottom*) ~ The 14-kilometre-long, dark core of Halley's Comet on 14 March 1986 photographed with the Halley Multicolour Camera aboard the European space probe Giotto. The mosaic, on which active geisers of gas and dust are visible, is compiled from 68 separate photographs. ► ESA/H.U. KELLER (MAX-PLANCK-INSTITUT FÜR AERONOMIE)

Page 77 ~ Close-ups of three planetoids. *Left to right*: (433) Eros (diameter at widest point approx. 33 km), photographed by the American space probe NEAR–Shoemaker; (951) Gaspra (58 km), and (243) Ida (56 km), both photographed from the American Galileo unmanned spacecraft. ► NASA/JOHNS HOPKINS UNIVERSITY APPLIED PHYSICS LABORATORY (Eros); NASA/JPL (Gaspra/Ida)

Page 78 ~ The bright A-ring of the planet Saturn with, at the top, the dark Encke Division, on 23 August 1981 photographed by the American Voyager 2 spacecraft from a distance of 2.8 million kilometres. ► NASA/JPL

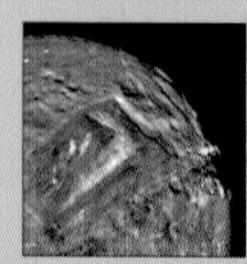

Page 80 (*top*) ~ Miranda, the small, battered moon of Uranus, photographed on 24 January 1986 by the American Voyager 2 spacecraft. The image is compiled from nine separate photographs. Miranda has a diameter of 480 km. ► NASA/JPL

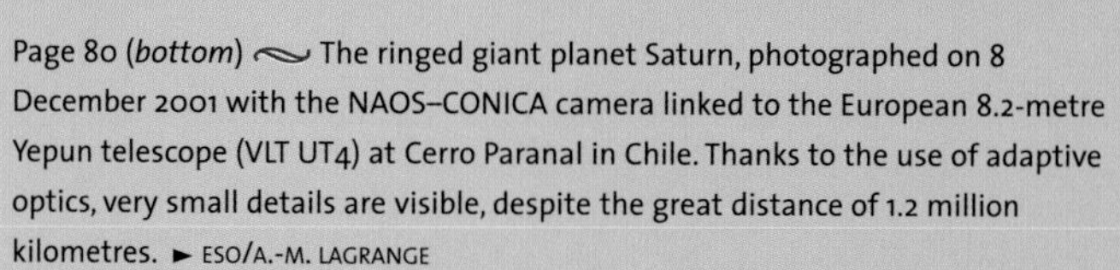

Page 80 (*bottom*) ~ The ringed giant planet Saturn, photographed on 8 December 2001 with the NAOS–CONICA camera linked to the European 8.2-metre Yepun telescope (VLT UT4) at Cerro Paranal in Chile. Thanks to the use of adaptive optics, very small details are visible, despite the great distance of 1.2 million kilometres. ► ESO/A.-M. LAGRANGE

Page 81 ~ In the early days of the Solar System, newly born planets including the Earth and Moon were subjected to heavy bombardment by cosmic projectiles. ► M.A. GARLICK, HTTP://WWW.MARKGARLICK.COM

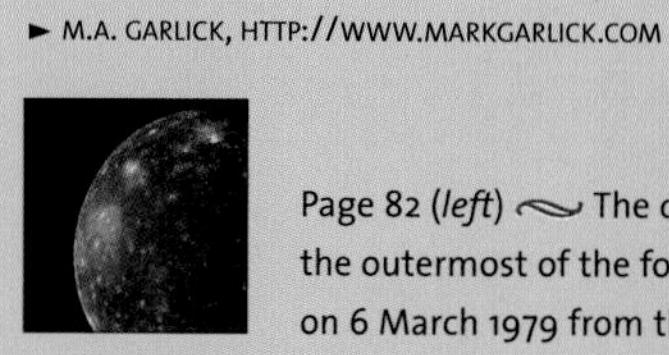

Page 82 (*left*) ~ The colossal impact crater Valhalla on Callisto, the outermost of the four large moons of Jupiter, photographed on 6 March 1979 from the American Voyager 1 spacecraft. Callisto

has a diameter of 4800 km; the concentric rings extend to almost 1,500 km from the centre. ► NASA/JPL

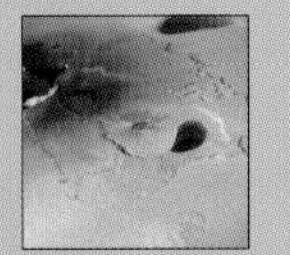

Page 82 (*right*) ~ Lava lakes and sulphur deposits around Tvashtar Catena, a chain of active volcanic calderas on Io, the innermost of the four large moons of Jupiter, photographed on 22 February 2000 from the American Galileo space probe. The area on the photograph measures approx. 250 × 250 km. ► NASA/JPL

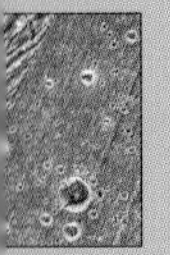

Page 83 (*top*) ~ Close-up of Nicholson Regio and Arbela Sulcus – ridges and parallel grooves on the battered surface of Ganymede, the second outermost of the four large moons of Jupiter, photographed on 20 May 2000 by the American space probe Galileo from a distance of 3350 km. The area in the photograph measures approx. 85 × 20 km. ► NASA/JPL

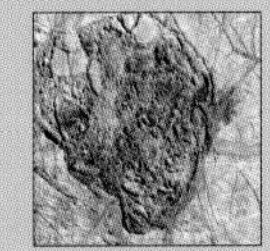

Page 83 (*bottom*) ~ Thera is a dark area on the bright, icy surface of Europa, the second innermost of the four large moons of Jupiter. This has probably been caused by material forcing itself upwards through Europa's mantle. The photograph was taken on 26 September 1998 by the American space probe Galileo. The area in the photograph measures approx. 240 × 85 km. ► NASA/JPL

Page 85 ~ A computer-generated artist's impression of Melas Chasma, a canyon hundreds of kilometres long slightly to the south of the Martian equator, depicting the distant past when there was possibly still water on Mars' surface. This three-dimensional image is based on detailed height measurements made by the Mars Global Surveyor. ► K. VEENENBOS, HTTP://WWW.SPACE4CASE.COM

Page 86 ~ The Moon setting on 14 July 1995 photographed from the space shuttle Discovery. ► NASA/JSC

Page 87 (*top*) ~ A three-dimensional image of Alpha Regio, an indented plateau on the planet Venus with a diameter of approx. 1300 km. To the bottom left of Alpha Regio is the large ovoid Eve. The vertical relief has been enlarged 23 times; the colours have been applied artificially. This image is based on radar observations from the American space probe Magellan, carried out in 1989 and 1990. ► NASA/JPL

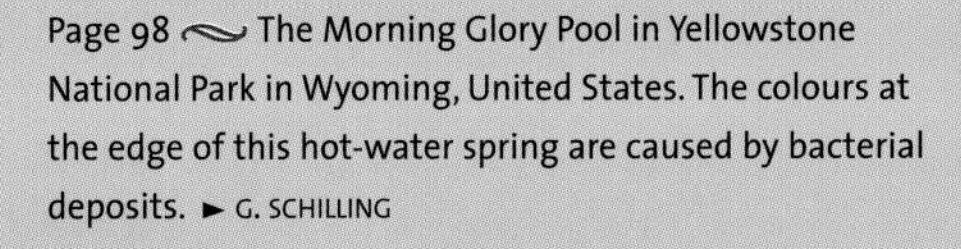

Page 87 (*bottom*) ~ The inner wall of the 7-km-wide crater at the bottom of the Newton Crater on the planet Mars shows flow patterns and narrow gullies that suggest that water flowed there fairly recently. Perhaps this water came from a layer of ice at some depth below the surface. This artificially coloured photograph is a mosaic of three photographs taken between January and May 2000 by the space probe Mars Global Surveyor. ► NASA/JPL/MSSS

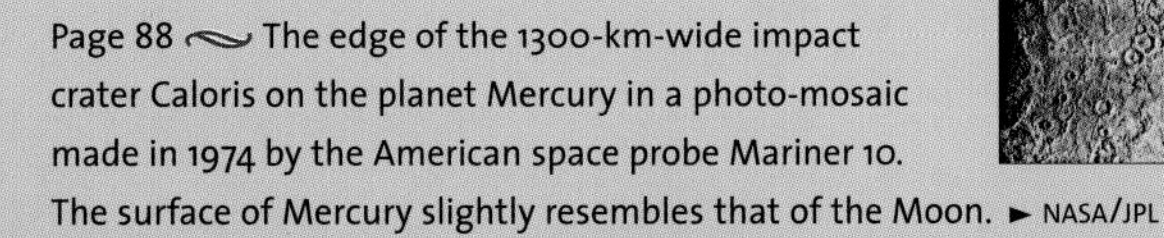

Page 88 ~ The edge of the 1300-km-wide impact crater Caloris on the planet Mercury in a photo-mosaic made in 1974 by the American space probe Mariner 10. The surface of Mercury slightly resembles that of the Moon. ► NASA/JPL

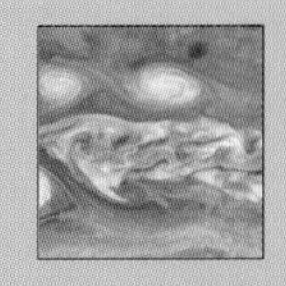

Page 89 ~ A mosaic of photographs of the atmosphere of Jupiter, taken in near-infrared by the American space probe Galileo. High, thin clouds are light blue; high, thick clouds are white; lower clouds are red. The area on the image measures approx. 35 000 × 16 000 km. ► NASA/JPL

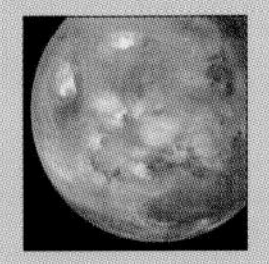

Page 90 ~ Thin, blue-white clouds of water crystals hang above the volcano summits on the Tharsis ridge on the planet Mars in this mosaic of photographs taken by the American space probe Mars Global Surveyor in April 1999. On the left is Olympus Mons, the highest volcano in the Solar System. ► NASA/JPL/MSSS

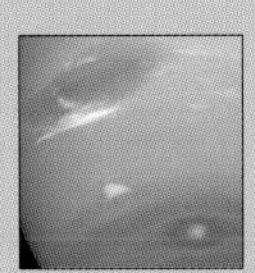

Page 91 ~ The blue, giant planet Neptune, photographed in August 1989 by the American space probe Voyager 2. The blue colour of Neptune's atmosphere is caused by the fact that methane gas in the atmosphere absorbs mostly red light. There are darker patches in the atmosphere and high, thin strands of cloud. ► NASA/JPL

Page 92 ~ The Earth is a planet of water, as can be seen from this photograph of clouds above the Indian Ocean, taken on 1 June 1999 from the space shuttle Discovery. ► NASA/JSC

Page 94 ~ The infrared-emitting nebula IRAS 16362-4845 shows up brightly against the dark background of the RCW108 complex, a molecular cloud at a distance of 4000 light-years from Earth in the constellation Ara (The Altar). The photograph was taken with the SOFI camera linked to the European 3.6-metre New Technology Telescope at La Silla in Chile. ► ESO/F. COMERON

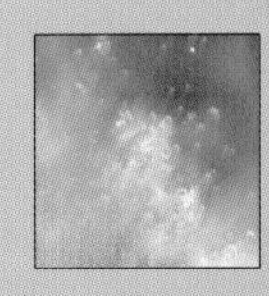

Page 95 ~ Small, comet-like concentrations of gas and dust in the Helix Nebula, a 10 000 year-old planetary nebula 450 light-years distant in the constellation Aquarius, photographed with the WFPC2 camera aboard the Hubble Space Telescope. ► C.R. O'DELL AND K.P. HANDRON (RICE UNIVERSITY)/NASA

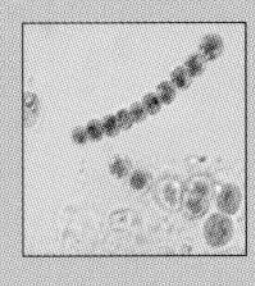

Page 96 ~ Micro-photograph of *Anabena* bacteria, a micro-organism that produces, among other things, nitrogen. ► CORNELL UNIVERSITY/W. GHIORSE

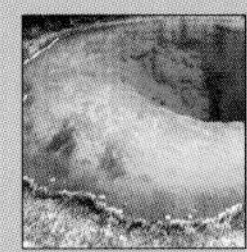

Page 98 ~ The Morning Glory Pool in Yellowstone National Park in Wyoming, United States. The colours at the edge of this hot-water spring are caused by bacterial deposits. ► G. SCHILLING

Page 99 ~ Artist's impression of the impact of a planetoid that ended the rule of the dinosaurs 65 million years ago. ► M.A. GARLICK, HTTP://WWW.MARKGARLICK.COM

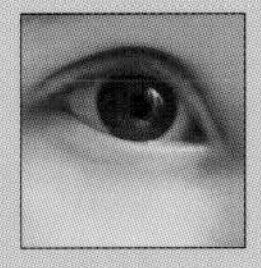

Page 100 ~ A unique re-ordering of a relatively small number of atoms, formed in the interiors of stars. ► PHOTODISC

Page 101 ~ Sunset at Stonehenge in Wiltshire, England. ► G. SCHILLING

Pages 102/103 ~ The Earth at night. The human race has changed the appearance of its planet radically. This map is based on observations made by the American Defense Meteorological Satellite Program. ► C. MAYHEW AND R. SIMMON (NASA/GSFC)

Page 104 ~ The dome of the 8-metre Frederick C. Gillett Gemini Telescope at Mauna Kea, Hawaii. To the immediate left of the dome is the constellation Southern Cross; further to the left are the bright stars Alpha and Beta Centauri. To the right of the dome is the bright Carina Nebula. ► GEMINI OBSERVATORY

Page 105 ~ On 12 February 1984, space-shuttle astronaut Bruce McCandless took a spacewalk using the Manned Maneuvering Unit (MMU), at a distance of approx. 100 metres from the space shuttle Challenger. ► NASA/JSC

Page 106 ~ Artist's impression of the distant future of the Earth: a cold, lifeless, scorched planet lit by the weak rays of the Sun that has turned into a small white dwarf. ► M.A. GARLICK, HTTP://WWW.MARKGARLICK.COM

Page 108 ~ An eruption of Mount Etna, Sicily, on 30 October 2002, photographed by astronauts aboard the International Space Station. ► NASA/JSC

Pages 108/109 ~ A sandstone formation eroded by wind. ► PHOTODISC

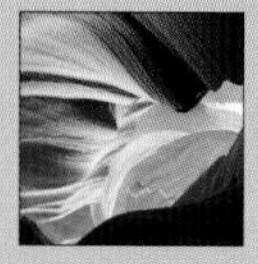

Page 110 ~ A detailed photograph of a group of sunspots taken with the 1-metre Swedish Solar Telescope at the Roque de los Muchachos observatory on La Palma, the Canary Islands. Thanks to the use of adaptive optics, details only 75 km in size are visible. ► ROYAL SWEDISH ACADEMY OF SCIENCES

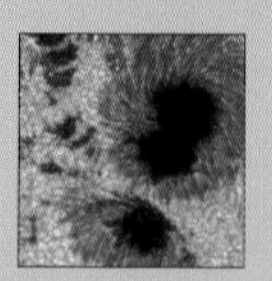

Page 111 (*left*) ~ A coronal mass-ejection (CME) of the Sun, taken in October 2002 with the LASCO camera aboard the American/European SOHO satellite. The bright surface of the Sun has been covered; the white circle shows the size of the Sun. Apart from the CME (left) so-called *streamers* in the corona are also visible. ► SOHO/LASCO (ESA/NASA)

Page 111 (*right*) ~ Artist's impression of the surface of a planet in an orbit around a red-giant star. ► H. FERGUSON (STSCI)/N. TANVIR (UNIVERSITY OF CAMBRIDGE)/T. VON HIPPEL (UNIVERSITY OF WISCONSIN)/NASA

Page 112 ~ The central part of the Halter nebula (M27), a planetary nebula at a distance of 1200 light-years in the constellation Vulpecula (The Fox), photographed with the 3.5-metre WIYN telescope at the Kitt Peak observatory in Arizona. ► N. SHARP, R. REED, D. DRYDEN, D. MILLS, D. WILLIAMS, C. CORSON, R. LYNDS AND A. DEY (NOAO/WIYN/NSF)

Page 113 (*top*) ~ The Little Ghost Nebula (NGC 6369), a planetary nebula approximately 3500 light-years distant in the constellation Ophiuchus (The Serpent Holder), photographed with the WFPC2 camera aboard the Hubble Space Telescope. ► NASA/HUBBLE HERITAGE TEAM (AURA/STSCI)

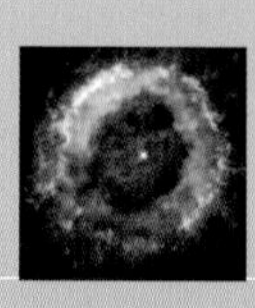

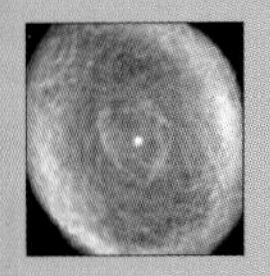

Page 113 (*centre*) ~ The Spirograph Nebula (IC 418), a planetary nebula approx. 2000 light-years away in the constellation Lepus (The Hare), photographed with the WFPC2 camera aboard the Hubble Space Telescope. ► R. SAHAI (JPL)/A.R. HAJIAN (USNO)/NASA/HUBBLE HERITAGE TEAM (AURA/STSCI)

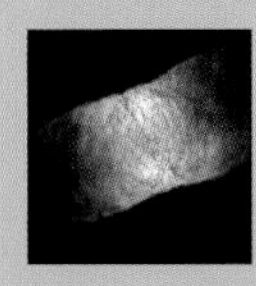

Page 113 (*bottom*) ~ IC 4406, a bipolar planetary nebula approx. 5000 light-years away in the constellation Lupus (The Wolf), photographed with the VLT Test Camera linked to the European 8.2-metre Antu telescope (VLT UT1) at Cerro Paranal in Chile. ► ESO

Page 114 ~ Artist's impression of the surface of Saturn's moon Titan. The composition of Titan's atmosphere is comparable to that of the primitive Earth. ► ESA

Page 117 ~ Artist's impression of Nova Aquilae, that erupted four times in the year 2000. Novae are thermonuclear explosions on the surface of a white dwarf (right) in a double-star system in which matter is transferred from one star to the other. ► CXC/M. WEISS

Page 118 ~ NGC 1232, a splendid spiral galaxy 100 million light-years distant in the constellation Eridanus, photographed with the FORS1 camera linked to the European 8.2-metre Antu telescope (VLT UT1) at Cerro Paranal in Chile. ► ESO

Page 121 ~ The Antennae (NGC 4038/4039), two colliding galaxies approx. 65 million light-years distant in the constellation Coma Berenices (The Hair of Berenice), photographed with the VIMOS camera linked to the European 8.2-metre Melipal telescope (VLT UT3) at Cerro Paranal in Chile. Many new stars and star clusters (blue in the photo) have been created as a result of the collision. ► ESO

Page 122 ~ Part of the Hubble Deep Field North, a small area of sky in the constellation Ursa Major (The Great Bear), that was exposed for tens of hours by the WFPC2 camera aboard the Hubble Space Telescope. The arrow marks the position of the most distant known supernova SN1977ff. ► NASA/A. RIESS (STSCI)

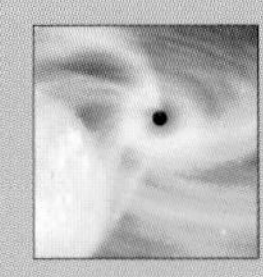

Page 124 ~ Artist's impression of a black hole in a double-star system. Matter from the accompanying star (left) is being sucked into the black hole, to disappear forever behind the event horizon. ► G. PÉREZ DÍAZ (IAC)

Page 125 ~ Artist's impression of a black hole as a deformation of (and possible escape from) four-dimensional space-time. ► M.A. GARLICK, HTTP://WWW.MARKGARLICK.COM

Page 126 ~ The Andromeda galaxy (M31), 2.2 million light-years distant in the constellation Andromeda, photographed with the 0.9-metre telescope at the Kitt Peak observatory in Arizona. ► T.A. RECTOR AND B.A. WOLPA (NOAO/AURA/NSF)

Index

Illustrations page numbers in italic